CIVILIZATION'S
TEMPERATURE

CIVILIZATION'S
TEMPERATURE

Effect of Climate on Humankind's History

Alexander Nikonov

Translated from Russian by

Paul R. Friedman

To order additional copies of this book, contact:
Xlibris Corporation
1-888-795-4274
www.Xlibris.com
Orders@Xlibris.com
73002

CONTENTS

All the historical and natural science data quoted in this book were provided by the Laboratory of Global Energy Problems and by its head, Professor Vladimir Viktorovich Klimenko, Ph.D.

In his new book, Alexander Nikonov speaks about shockingly interesting studies by V. Klimenko who did enormous work on researching the climate in various regions of the Earth. V. Klimenko integrated all possible methods of studying the weather into a single system and got very impressive results.

Having read this book, you'll be able to carry on endless talks about weather no worse than any Brit.

This author states with all due responsibility that this book could not have come into being if it were not for climatologist Vladimir Klimenko, who is in fact the full-fledged coauthor, because with his studies of many years, he laid the foundation of this work, built its walls, and put a roof over it. All that was left for me was to take a paintbrush and to paint the building in the public-favored colors, and in addition, perhaps, to willfully cut an entrance on the other side. Or rather an exit, because I was not in full agreement with Klimenko's obviously pessimistic conclusions concerning the civilization's future.

The Author

What are humans indebted to for their success? What human efforts is our life indebted to for its arrangement? What, in the human sense, depends on humans, on all of us? What do humans depend on?

Sever Gansovsky

When ground was covered with first snow,
The child awoke from chilly blow,
And to stay warm in dark nightlong
He took the harp and sang his song.

Dmitry Bavykin

FROM THE AUTHOR

"I want to go on a cruise," said my wife, rolling her eyes and being lost in a reverie.

I did not respond.

It happens with her. The last couple of years she developed a liking for sailing in the waters. It all started slowly, with cruises along the Volga, but ended . . .

"I want to sail along the Nile! Or Europe. Or Danube, for example. The 'Danube waves,' charming . . . I saw some ads—it could be in winter. It's cheaper in winter. Not on the New Year's, of course, when it's high season, but later—in January-February. And then, at this time, several countries could be visited, Vienna, Budapest . . ."

Indeed, it's cheaper in winter. Because it's cold, dank. Kind of not nice. Gray sky. It's no warmer than about five to ten degrees above zero (40-50°F). Maybe even a whole zero (32°F)! Cold rain . . . I was in Bratislava one winter. The leaden Danube was rolling majestically its unsightly waters, and rocking slightly on them was a floating restaurant where my friends and I were celebrating the coming of the New Year, new century, and new millennium.

At the turn of millennium, the flow of time is felt kind a sharper, for it runs away from under the feet like the Danube water. And this water did not amaze anyone, neither in the new century nor in the one just passed. But in the past, a picture like that unfrozen Danube could not be even imagined. Yet nowadays, a winter cruise along European rivers is in the order of things, and navigation does not stop here: Europe is famed for its warm winters. There is even a term for this—"Euro-winter." Euro-winter means rain, green grass, light jackets . . . The Russians who happened to be in wintry Europe are telling one another, with a note of enthusiastic amazement, that in January-February in Montenegro, the first flowers start blooming with a stupefying smell; in January in Barcelona they were walking in the city with only their shirts on, because the sun warmed up the air to plus eighteen degrees (65°F).

But it was not always like this. And it will not always be . . .

LIKE PONTIUS OVIDIUS ICE

Instead of Introduction, Peroration, Penetration, Domination, Excitation . . .

> *The soul freezes, the soul toughens, the soul hardens, like the Pontius Ovidius ice at the shores of Ukraine. The Danube freezes, the fields toughen, the firmament hardens, and the snows of the Black Sea, gloomy outskirts cover the eroded Scythian barrow that became your grave.*
>
> *Sergey Zavyalov*

It's hard to say why Ovid was exiled. Historians argue about it even today. Some think it was for pornography, others, for flirting with emperor's granddaughter. As you have realized, he was the same type as the great Russian poet, Pushkin, this Ancient Roman poet, Ovid, quite a mischief-maker.

Publius Ovidius Naso was born in 43 BC to a rich equestrian family. As the custom for a noble offspring was, the lad got decent education in Rome. He tried to work as a bureaucrat, but had distaste for the job—Ovid was more attracted to poetry. Thank God, the means allowed to do nothing and to enjoy creating doggerel.

Soon, the young man became famous all over Rome for his frivolous poems, which in some of their parts turned into outright pornography. But Ovid managed to shroud his base porn in such pretty forms that . . . well, what's there to be said—talent is talent!

Nevertheless, the chaste emperor Augustus took a serious dislike to Ovid for his jolly frivolities. In all times, life is tough for life-loving poets. Besides, Ovid's private life was also not too successful: a divorce, then the second one. He found some calm only in his third marriage.

However, the heat in blood fades away with years . . . Gradually Ovid became calmer and started writing not just porn verses, but some serious works as well. They were the ones that brought him worldwide fame. Soon he became the most beloved poet of Rome. But not of the emperor! Nationwide love for Ovid did not prevent Augustus from exiling the genius to Siberia. Siberia, for Romans, was the area where the Russian rulers later exiled their subjects to for the purpose of health improvement—the Black Sea shores.

. . . Oh, what places! Crimea, Taman, subtropical Caucasus, golden sands of Bulgaria. Slightly to the north of the golden Bulgarian sands are no less golden sands of Romanian resorts Mamaia and Constanta. Lands of plenty grapes! By the way, it was Constanta where unfortunate Ovidius Naso was exiled to. Although the city bore a different name then, Tomi, but that's irrelevant . . . Having arrived in the place of his exile, proud Ovid started immediately sending flattering letters to the emperor in Rome, in which he humbly begged to get him out of there. So what did the genius of Roman philology dislike at the resort?

He wrote piercing verses about it. I shall quote them in their entirety: they're worth it. And you should pay attention to the specifics of the climate in the resort places. But before we start, one comment has to be made: in those times, the Danube was called the Istros.

So,
Siquis adhuc istic meminit Nasonis adempti,
Et superest sine me nomen in Urbe meum,
Suppositum stellis numquam tangentibus aequor
Me sciat in media vivere barbaria . . .
(Hereinafter, just the translation) (*English translation is done from the Russian—Translator's note)*

If someone still remembers Naso, taken away from you,
And my name still lives in the City without me,
Let them know that I live under constellations
That never saw the sea, amid barbarism.

———

16

I'm surrounded by savage tribes of Sauromatae, Bessi, and Getae,
Oh, these names are so unworthy of my talent!
And still, while it's warm, we're protected by the waters of Istros,
It averts the wars by the flow of its waters.

But when the gloomy winter shows its stiffened face
And the ground becomes white from marmoreal ice,
When Boreas and snow prevent from living under Arcturus,
Then it's clear that these tribes are depressed by the Pole of Frost.

Snow lies everywhere, and to prevent sun and rains from melting it,
Boreas fortifies it and makes it eternal.
Thus, before the old snow melts, the new one falls,
And in many places, it usually stays two years in a row.

And such is the force of raging Aqualon
That he razes high towers to the ground and blows away the roofs.
People protect themselves from bitter cold with animal skins and sown
      pants,
The only open part of their body is the face.

When in motion, the hair often rings from icicles hanging on it,
And the white beard shines, being covered with rime.
The wine taken out of container stands and retains its shape,
And it is served not by drinks but by pieces.

What now? Should I narrate, how being frozen over,
The creeks harden and brittle waters are cut out from the lake?
Even Istros, which is as narrow as a papyrus-carrying river
And which flows with its many estuaries into the huge sea,

Freezes over from the winds that harden its blue waters
And slowly pulls its invisible waters to the sea.
And in places where ships used to travel, now people's feet travel
And horses' hooves trample the waves hardened by the freeze.

—

And on new bridges, on top of rolling waves,
Sauromatian bulls drag barbarian carts.
My story is hardly believable, but since nothing is gained by deceiving,
My evidence should be perceived as completely trustworthy:

I saw how the huge sea was congealed under ice
And the smooth cover froze over the still waters.
And I not only saw this: I stepped on the water's hard glassy surface
And I stood over the waves with my feet staying dry.

If you, Leander, had a similar sea in your time,
Then the narrow strait would not have caused your death.
Dolphins, bending, cannot jump up into the air:
Severe bad weather holds back their attempts.

And let Boreas hoot and wave his wings,
He will not cause any rough waters in the frozen deep,
The ships captured by the hard freeze will stay in the marmoreal,
And the oars will not be able to cleave the hardened waters.

I saw how the fish, as if tied up, froze in the ice,
But part of them remained alive even then.
And as soon as the wild force of furious Boreas
Will pack the waters of either the sea or the flowing rivers,

Right then, barbarian enemy on fast horse
Crosses the Istros, smooth from severe Aqualon.
The enemy, whose strength lies with his horse and far-flying arrow,
Destroys vast areas of neighboring lands.

Some dwellers scatter, and since no one guards the fields,
The enemies pilfer the unguarded properties,
The wretched villagers' properties: cattle, squeaky wagons,
And other riches of poor villagers.

Some others are taken prisoners with hands tied up behind their backs,
While looking back in vain at their village and house.
Yet others, poor souls, fall to the ground, pierced by hooked arrows,
Since the flying iron is impregnated with poison.

Whatever the enemies can't take away, they destroy,
And the barbaric flames burn down the innocent shacks.
Even in the times of peace, the dwellers shudder being scared of a war,
And no one is plying the plow to till the land.

This country either sees the enemy or fears not seeing it,
And the neglected land stays aimlessly idle in torpid futility.
Here, a sweet bunch of grapes does not hide in the shade of leaves
And the bubbling new wine does not fill the deep vats.

The country is devoid of fruits, and Acontius would have nothing
On what to write his words so his beloved could read them.
You see here only bare fields with no vegetation and no trees . . .
Oh, what places, where a happy person should never come!

And so, although the huge world is widely spread out,
This particular land is chosen for my exile!

*   *   *

So, "in that steppe far-off, coachman froze to death." (*Quote from an old Russian song—Translator's note.*) You would agree it was no time for steamship cruises along the Danube. Back of beyond! As to the wine mentioned by Ovid, eaten by pieces broken off from a red ice block, it was imported. The reason is that in a climate resembling that of present-day Siberia, viticulture was out of the question. And archaeological excavations confirm this conclusion: in the strata corresponding to this time, practically no grape seeds are found. Grapes did not grow ripe then in Bulgaria and Romania. Nor did they grow ripe in Crimea.

Have you noticed, in how much detail Ovid describes ice—hardened water on which one can walk like on dry land? At that time, this was amazing to residents of the Eternal City. And generally, such severe frosts were unusual then. Two hundred years before Ovid, famous scientist, Eratosthenes of Cyrene in his work *Geography* adduces an interesting fact. Residents of the city of Kerch (it was then called Panticapaeum) have sent to their capital a copper vessel that had burst from the freeze and had the engraving: "If someone doesn't believe what's going on here, let him be convinced by looking at this hydria that had been submitted by priest Strotius not as a beautiful tribute to God, but as proof of severity of our winter."

However, it would be wrong to say that residents of the Eternal City had absolutely no knowledge of ice. It was only to the great poet's contemporaries that ice, perhaps, was something unusual. But four hundred years before Ovid, Rome also saw that neither grapes nor olives could grow ripe—it was cold indeed. In winter, the Tiber froze over and the snow covered Italian fields up to forty days a year.

Grapes were not cultivated in Northern and Central Italy up to the end of the third century BC, although Italian peasants, of course, were well acquainted with grapevines—grapes were first brought to the Apennines around 700 BC. Greek colonists, who founded the city of Cumae (close to present-day Naples), were involved in viticulture, but in the next five hundred years of colonization viticulture never moved north of southern Italy. Some old Greek texts were found that don't recommend peasants to cultivate grapes and olives, north of Naples—they would freeze . . . Only in the third to second centuries BC, when the climate warmed up, grapes and olives spread all around the entire Boot.

Professor Vladimir Klimenko, one of the most competent modern climatologists, once told me, shaking his head, "Over the human history, climate changed more than once and catastrophically at times, but modern historians and especially popularizers simply ignore this because they don't know it. At times, some plain funny things happen. For example, I'm watching a BBC movie about the Exodus of Jews from Egypt, and I see the following scene: unfortunate Jews are crossing the desert on camels. But they had no camels then—camels were domesticated in Africa much later, only in the era of Julius Caesar. Yet even more important, there was no desert then in that place! The landscape was entirely different—Savanna was blooming there . . ."

One of the most fascinating scientific problems is to track the influence of climate on history of humanity. Scientists attempted it more than once, in order to learn how climate and geography affect fate of civilizations, character of nations, their moral and culture, but those were amateurish attempts. Only in the last decade, the large-scale reconstruction of climate was completed for the last ten thousand years, and it became possible to superimpose the graph of climate fluctuations over human history. After that, it became possible to make some curious conclusions, which were also made by aforementioned Klimenko:

"Wherever we look on the globe, whichever era we draw our attention to, a natural phenomenon can be observed everywhere: cooling periods are the periods of greatest scientific and cultural breakthroughs, the periods of establishing great empires. And the warming periods are times of collapse, of empires, cultural stagnation. This rule works practically with no exceptions.

"Interesting. Especially when recalling that the collapse of the Soviet Empire happened in the time of global warming. Please continue, my friend . . ."

PART 1

Snail on a Slope

It's known since long ago that when a person is keen on something he would let nothing pass. Klimenko is a person. And due to his keenness, he is able to see what's imperceptible to others.

"Since I'm involved in climate, when I read Russian classics I subconsciously note everything related to weather. For instance, when reading Dostoyevsky, I noticed many times that his characters walk in summer in St. Petersburg in overcoats. In *The Brothers Karamazov* in one episode (taking place in late October), the author mentions casually that the frost was down to -10°C (+14°F). And it's clear from the context that this was a common phenomenon . . . Gilyarovsky has an episode describing his first visit to Moscow, to the present-day Komsomolskaya Square. This happened on October 19, 1876. Gilyarovsky narrates how having got off the train, with all his bags, he climbed over a snowdrift and got into a cab (horse-cab, that is). And he describes it very casually: what's the big deal, snowdrifts in Moscow in mid-October . . .

"As it turns out, in the nineteenth and early twentieth centuries, climate in Russia and on Earth was quite different, not the way it is today. But has it changed much? For the last hundred years, the annual average temperature in Russia has risen by approximately one degree Celsius (1.8°F). Some think it's not much. But this is by no means true. One hundred to one hundred fifty years ago, it was significantly colder in Russia. In Karelia, for instance, the forty-degree frosts (-40°F) were at times recorded in April! Today, this is something unthinkable . . . The present-day dwellers of communal flats complain in January about the rare forty-degree frosts, calling them 'unprecedented' . . . Back at the time of Nicholas I, *Journal of Ministry of the Interior* was first published in Russia. It would be more appropriate to call it the first meteorological almanac in Russia because it published most detailed reports on weather phenomena. One can learn from the *Journal* that even in the early twentieth century in Moscow snow fell heavily in June, and it was not something exceptional! The June snow fell even in Kiev province.

"Climate is an amazing thing."

CHAPTER 1

Thin Film of Life

One of the books on geology describes the following picture: imagine our planet as a 2.5-meter (100") diameter ball—a ball like that would fit exactly from floor to ceiling in an ordinary Moscow apartment. In this case, the world ocean with an average depth of 5 kilometers (3 miles) would be represented on this ball as a 1 millimeter (0.04") thick film. As to the atmosphere, if an airplane loses its pressure at an altitude of 10 kilometers (33,000 feet), everyone dies: it's impossible to breathe without an oxygen mask, that's how rarefied the air is. So, the 10-kilometer atmospheric ocean on our ball is equivalent to a thin layer of 2 millimeters (0.08").

Did you picture the Earth in the form of a ball that fits from floor to ceiling with a thin film of air and water on it? So, it's in this thinnest film that our entire life is concentrated, our entire history. What's also concentrated here, in this same film that's spread around our planet, is the entire climate.

Climate is also quite a subtle thing. During billions of years, many things happened on our planet: Earth was shaken by powerful volcanic eruptions; mountains sprang up 10 kilometers (6.2 miles) high on flat areas, which later became ocean bottom; the planet has survived more than one hundred asteroid attacks similar to the one that killed dinosaurs seventy million years ago. But in spite of all these catastrophes, the average global temperature never fell more than 8°C (14°F) or rose more than 10°C (18°F) with respect to what it is today.

All I want to say is that climate fluctuations, affecting history in a most dramatic way, usually don't exceed several degrees or even fractions of a degree! For instance, the Roman Empire (I shall discuss this later) was ruined by the drop of the global average temperature by mere half a degree (1°F). The climate crawls like a snail, hardly noticeable, and destroys civilizations.

Now you know what climate fluctuations are. But you still don't know what climate is . . .

A small town of Nicaea (currently Iznik), located not far from the present-day Istanbul, is known for having hosted in the year 325, the First Ecumenical Council of Christian Church. This historic event was to some extent a turning point in history of Christianity and thus made the tiny Nicaea famous. But Nicaea is also famous for the fact that in old days a scholar by the name Hipparchus dwelled here. It was this observant Greek who introduced the word *climate* into scientific vocabulary. In ancient Greek, *clime* is "slope." (The word *climax*, by the way, is of the same root.)

But what some uncertain slope has to do here and why Hipparchus called climate *climate*? As it was, the ancient scientist justly stated that climate depends on the "slope" angle, at which the sun beams fall on the surface of the Earth. Oh, what a keen-witted people they were, the Greeks! And they were curious, too: since the time when Greek civilization started, practically all the great Greeks wrote about climate—Aristotle, Plato, and Herodotus . . . As to Aristotle, he even left for progeny his classical work *Meteorology*; what a pity that only its fragments reached us.

Aristotle became the first scientist to point out that climate affects people's temperament and appearance, human behavioral stereotypes, and type of government structure. When pondering over a question why it was, they, the Greeks, who have become such a civilized nation, while northern and southern barbarians remained savages, Greek philosophers decided that responsibility for all this rests with climate. In northern latitudes, people cannot afford civilization—they are preoccupied with survival under severe conditions. Southern people cannot afford civilization either—food products in the land of plenty are so easily obtained that there is no need to improve skills and make inventions. And only in the middle latitudes where Greeks were fortunate to live, humans can (they have free time for that) and must (since not everything is easily obtained and the brains have to be used) think about the essence of things.

Climate was also of interest to Aristotle's disciple, Alexander the Great, of Macedon. His conquest campaign to the back of beyond could be called partially a scientific and research endeavor. Alexander wanted to see with his own eyes the river edging the back of beyond, called the Ocean, which he heard so much about from his teacher. Just as Napoleon later who went on his conquest campaign with a multitude of scientists, so did Alexander who was dragging along a crowd of scholars everywhere. However, upon arriving in India, not only his scholars, but also Alexander himself became terribly interested in a local climate phenomenon—the monsoons. For Europeans, it was astounding: the winds, which are known for their variability, in India blew for some reason with an automatic precision in summer from the ocean, in winter from the mountains, and never other way round.

It took another couple of thousand years plus, before the nature of this amazing phenomenon was explained, which shortly after was amazing no more. This happens with all sorts of tricks and religious miracles: as soon as the secret is revealed, comes understanding, slight disappointment, and search for new secrets. Unquenchable and indomitable is the human spirit in its curiosity—the main engine of progress . . .

Sir Gilbert Walker, the Oxford graduate, took a job in early twentieth century in India with Her Majesty's Royal Meteorological Service. Like all the other inquisitive Europeans, including Alexander the Great, he got interested in the monsoon phenomenon. But what distinguished Gilbert from other Europeans was that he was the first person ever to explain the monsoon's nature. It took Gilbert twenty years to solve the problem.

As we have mentioned, in summer the wind blows from the ocean onto the land. Why? Under the hot southern sun, the Hindustan Triangular warms up and heats the air at the ground level. The hot air is less dense and rises, causing it to be replaced by the more dense air from the sea. And this air is also more humid—it's from the sea, after all! It is these ocean winds, which brings the water necessary for agriculture—the rains, without which there would be no harvest.

In winter, everything is vice versa: the wind blows from the north. And here's why . . . In the north, the Hindustan peninsula is edged with Himalayas. And behind the Himalayan wall is the high-mountain Tibetan Plateau. It's very cold there in winter, with lots of snow and ice. The air gets very cold over there. And the cold air, as it is known, is denser than the warm air. And so it starts to flow down from the mountains to the Indian valleys, replacing the lighter warm air.

By the way, the same thing is inherent in another curious phenomenon called *bora*. Bora happens often, for instance, in the city of Novorossiysk

—

and brings lots of problems to residents and seafarers. Bora is a nasty cold wintry wind blowing from the mountains. Imagine the warm southern sea. Not far from the shore is a mountain massif that protects the shore from the cold air to the north. But opposite Novorossiysk, mountains are slightly lower—it seems like there's a gap in the mountain range, as if a fairy-tale giant bit, off an edge of the mountain ridge. And if the mass of cold air accumulated behind the mountains becomes suddenly so huge that it reaches the mountain-ridge top, the cold air starts flowing down through the gap, filling up the city. This flow of cold air can flow nonstop many days at a time, because the ocean of cold air masses is huge and the "hole" is small. And at this point, all the ships that are in the port or in the roads have their masts covered with a very thick (many centimeters or inches) coat of ice. Ice covers wires, riggings, and tree branches . . . Iced-over yachts could even overturn, because their ice-covered heavy masts could outweigh the keel.

By the way, the word *bora* stems from the name of ancient-Greek god of wind, Boreas, who appeared in the works of our sufferer Ovid . . . But let's get back to climate. As modern science sees it, climate implies averaged weathers for the latest thirty-year period. Today, the contemporary climate model is believed to be the period from 1961 through 1990, although strictly speaking, the climate on our planet has changed mightily since then. And it has changed for the better. Some time ago, Gilyarovsky had to walk with his bags over snowdrifts, and it was in mid-October. I'm writing these lines in early December, and I can see through my window that the ground is black, and the window is wet from raindrops. The showers keep drizzling for almost the whole day and have no intention of turning into snow. In this era, we have a problem with skiing: when going to the mountains for the New Year, we never know whether or not, we'll find snow up there. Cold sweat.

Chapter 2

Climate Orchestra

"There's a four-story house, each floor has eight windows, on the roof there are two dormer windows and two chimneys, and there are two tenants on each floor. And now, gentlemen, please tell me, in what year did the doorkeeper's grandmother pass away?" This problem was given by the good soldier Svejk to the medical commission that was checking his psychiatric health using the Kallerson and Weyking system.

The problem is no doubt interesting, but insoluble for obvious reasons: it lacks data necessary to solve it, while the data available are worthless. The problems of weather and climate forecast have a similar nature. Climate kitchen, which seethes in the thinnest (compared to the planet's size) layer of hydrosphere and atmosphere, is perhaps one of the most complex processes known to humans. Its computation is a task of unimaginable difficulty. The system is three-dimensional, three-phase (liquid, solid, and gas), and most importantly, it depends on tons of internal and external parameters that are not always obvious . . . Besides, any problem could be solved only if it's formulated correctly. But how to formulate it correctly? What does the climate depend on? Which data need to be entered in the problem, which are redundant, and which could be ignored for simplification?

Solar radiation . . . volcanic eruptions . . . total number of sheep on the planet . . . economic development level, and the area of polar caps . . .

Which of the above affect the climate? Which of these parameters are superfluous? Besides those listed, are there any other "influential" parameters? I'm responding in the order of the questions: all, none, yes.

Let's start with the most important: Sun. It's obvious that it's the main culprit of all the weather and climate disgrace happening on Earth. If it were not for this yellow dwarf, the entire atmosphere of the planet, along with climate and weather, would simply fall onto the cold surface in the form of dry snowy flakes of frozen gas.

The Sun never stops surprising the researchers with new tricks. For example, just couple of decades ago, scientists believed that the amount of heat coming from the Sun to Earth decreases during periods when the Sun has the most number of spots (the period of elevated "spottiness" of our luminary is called the period of solar activity). This was a reasonably logical assumption: dark spots are the zones on the solar surface where temperature is lower than around them, causing them actually to look dark against the background of the beaming space. (Temperature of sunspots is almost half of that of the "working" solar surface.) And since there are many dark spots, then the average temperature of the solar surface is lower, as is the heat flow . . . And only recent studies proved the opposite.

The questions whether the shine of the Sun is steady or fluctuant were puzzling scientists for a long time. More than one hundred years ago in the mountains of the United States, observatories were built to measure the intensity of heat radiation emitted by the Sun. But at that time, attempts to measure minute fluctuations of the heat flow have failed. They were discovered just recently, during the era of extra-atmospheric research: in December 1978, an American satellite, Nimbus-7, was launched. This outer-space laborer flew twenty long years around the Earth, watching our luminary; it determined that solar radiation not only fluctuates, but follows closely the solar activity—the Sun emits more heat when it has many spots, and vice versa. But how much more?

The difference between maximum and minimum heat flows is approximately 2.5 W/m^2, or 0.15 percent of the "norm." Pennies! But these pennies seem to be enough to change the average global temperature by several tenths of a degree or even by a whole degree Celsius (almost 2°F). Is it insignificant? It's sufficient to shake up social foundations, which will be convincingly shown later.

The fact, that the Sun periodically "gets sick," being covered with dark spots, was known to humans even before invention of telescope. Chinese discovered the spots on the Sun, back in 165 BC—the spots could be seen then by the naked eye. And not only then! Periods of super-high solar activity happened later, too. For example, Russian chronicles reported a similar phenomenon in 1372. With the telescope invention, observation of sunspots became the matter of technique, so to speak.

It's funny, by the way, how things developed after this invention. In Soviet encyclopedias, you can find a statement that telescope was invented by Galileo. Italian encyclopedias do not dispute this for obvious reasons. But Germans have different opinions—German encyclopedias give two versions: 1) telescope was invented by Christoph Scheiner from Ingolstadt; 2) telescope was invented by David Fabricius from Resterhafe, which is in Eastern Friesland. British encyclopedias believe that telescope was invented by Thomas Harriot—naturally, an Englishman. But this Harriot, upon inventing telescope and honestly describing his invention, cast aside his invention into his desk, never mentioning it to anyone. It happens this way with Brits. For instance, an Englishman Cavendish discovered argon, but without ever mentioning it to anyone he through his notes into his desk, which resulted in the second discovery of argon about one hundred years later, with the "discoverer's" fame belonging to Ramsay.

But no matter who invented the telescope, with this instrument, things started looking up. In the early seventeenth century, dark spots on the Sun were observed by Galileo Galilei, Thomas Harriot, Christoph Scheiner, and David Fabricius ... Later, a German amateur astronomer, Heinrich Schwabe discovered the famous eleven-year cyclic occurrence of sunspots: every eleven years, for a reason unknown, the number of sunspots sharply increases. But again, not always! Occasionally, the Sun suddenly seemed to forget that it should periodically increase its activity while getting covered with spots, and came to a standstill with its tranquil, steady burning. This happened, for instance, during the Maunder Minimum, named after English astronomer Maunder. During that entire period of seventy years (1645-1715), the total number of observed sunspots was fifty. It was one thousand times less than usual! The Sun was utterly inactive. This was a hard time for our planet. Average winter temperature dropped one to one and a half degrees (2-3°F), while the average global temperature dropped half a degree (1°F).

It would seem that in the global scale of things it's not much—half a degree (1°F). And even one and a half (3°F)! So what, if the temperature on a winter night is -16.5 degrees (+2°F) instead of -15 (+5°F)? Not a big difference! However, the climate system is so subtly adjusted, that the drop in average global temperature by half a degree (1°F) causes big weather troubles "at locales." By the way, half a degree (1°F) is exactly how much the average global temperature rose in the last century. And this rise made everyone talk about global warming, glaciers melting, the World Flood . . . At the same time, during the Maunder Minimum, it would be appropriate to talk about global cooling—those were the coldest decades in the last two thousand years.

This was the Stradivari time. In a specialized scientific magazine *Dendrochronology*, an article published by two Americans showed that the violins of the famous Italian master are indebted for their majestic sounding specifically to global cooling. Antonio Stradivari was born right at the start of the Maunder Minimum, and he created his most valuable instruments from 1700 through 1720. The first who drew attention to this coincidence was the climatologist from Columbia University, Mr. Lloyd Burkle. He decided to check how the Maunder cooling affected alpine spruces from which Stradivari made his violins. He contacted dendrologist Henri Grissino-Mayer. It turned out that in the five-hundred-year history of alpine conifer forests, there was a period of extremely slow growth. Did you figure it out? Indeed, it happened exactly during the Maunder Minimum. Everywhere—from France to Germany—from 1625 through 1720, spruces, pines, and larches were growing with enormous efforts; the annual rings seen on saw cuts are very compact and narrow. "The compact wood sounds better," was the climatologists' conclusion. Thus, the severe climatic conditions had a beneficial impact on arts.

. . . Long winters, cool summers. We know where such conditions are today. If Stradivari were alive, he would not be making his violins from Italian pines; the old man would have all the reason to order Russian lumber, made in Siberia.

Stradivari was lucky—he lived in Italy. Much less lucky were those who lived a bit to the north. The worst hit by Maunder cooling were Finland, Sweden, Russia, Norway, and Estonia—a series of lean years mowed down up to 50 percent of the population. Those days, Russia was ruled by Peter I, "the Great." From his notebooks, we learned about climatic anomalies of that era: the first siege of Azov failed because on October 1, on the shores of the Azov Sea the Russian troops got buried under the snow. Today this is the time when the "velvet" season at the Black Sea shores is about to close.

Getting ahead of my story, we can say that the new twenty-first century is the time of a deep and prolonged solar minimum similar to the Maunder Minimum. This minimum should come no later than the twenty-fifth solar cycle. Currently, we are in the twenty-third. (Following the idea of Bern Observatory director Mr. Rudolf Wolf, after whom the Wolf numbers are named, the numbering of solar cycles is kept since the mid eighteenth century. It was at that time when the regular observations of sunspots started.) The twenty-third cycle of solar activity started in 1996 and should end in March 2008. The next cycle, the twenty-fourth, should end in 2020, which is when the twenty-fifth cycle starts. According to astronomers' forecasts, solar activity in this cycle will not exceed 50 units. For comparison, maximum in the twenty-second cycle was 155 units, and the absolute maximum of the twentieth century was 190 units, which happened in 1957. Anything less than 100 units is catastrophically small.

The solar activity minimum forecasted by scientists should last through the end of the century. As a result, intensity of solar radiation will decrease so significantly that the average global temperature will fall by the same half a degree (1°F) that killed half the population of Europe during Peter the Great era. And if you think that humans are better protected today from famine and crop failure than three hundred years ago, then you are profoundly mistaken. As Zoschenko wrote, "Stock up with coffins, bastards!" We can only hope that we won't have to fight global cooling instead of global warming. But we shall talk about the future in the last part of the book.

So, we have ascertained that climate is affected by fluctuations in solar activity. But the warming up of the Earth's surface depends not only on the Sun's activity but also on the location of the Earth itself—remember Hipparchus: *clime* means slope, angle. In other words, the warming depends on the angle, at which the planet turns its crown to the star. Since preschool years, everyone knows that the Earth rotates about the Sun and about its axis. The period of Earth's one revolution about its axis is 24 hours and about the Sun 365 days plus some change. But these two parameters are not the only characteristics of Earth's rotation.

The axis of rotation, which passes through the North and South poles, is inclined at an angle of 66.5° (or to be exact, 66°33') with respect to the plane of its orbit around the Sun, or 23.5° off the vertical. That is why the school globe is always "crooked," and you would never find anywhere a globe whose metal axis, crowned with a black plastic ball victoriously protruding from the North Pole, would be perpendicular to your desk. It's always tilted, because it reflects the existing reality of our planet's position.

While revolving about the Sun, the Earth puts its either North or South Pole toward it, thus causing the change of seasons: when more sunshine falls on the North Pole, summer comes to the Northern Hemisphere, and when more sunshine falls on the South Pole, the summer is in the Southern Hemisphere.

But 66.5 degrees is the average angle at which the axis of Earth's rotation is inclined with respect to the ecliptic plane. This angle constantly changes and drifts by one and a half to two degrees. This drifting is called nutation. When the angle is smaller, more heat gets to the "crowns," and the polar icecaps start melting. When, on the other hand, the angle increases, more heat falls onto the low, equatorial latitudes and less onto the polar areas. And then the polar icecaps expand. The period of nutation is forty-one thousand years.

The cycle of eccentricity (contraction-expansion of the orbit ellipse) is ninety-three thousand years.

And then there's such a thing as precession, which is cone-shaped oscillation of the Earth's rotation axis. Period of the precession oscillation is twenty-three thousand years . . . In a museum in the Belgian city of Lier, an amazing mechanical clock is on display. It's made by a clockmaker, Louis Zimmer, in 1935 and has several clock-faces. On one of the clock-faces, the world's slowest clock-hand moves—it makes one complete revolution in exactly twenty-three thousand years. When Einstein saw this Zimmer clock, he was very much impressed with it and warmly congratulated the master on creating such a cool gadget . . .

So, all the mentioned astronomical cycles are none other than cycles of climate oscillations. Astronomy is an exact science, like a clock. And not terribly complicated, by the way. Back in the old times of Ptolemy, it was possible to forecast solar eclipses for hundreds of years ahead. That's why it would seem that knowing with what accuracy the heavenly gears are running would make it easy to forecast climate changes. Alas! Instruments of climate orchestra sound cacophonously. Oscillations of different frequencies superimpose upon one another in a very complex way, like ripples running on high waves, creating a complex perturbation pattern. And if taking into account that it's not just astronomy affects the climate, the picture becomes tooth-crushing indeed.

One of the very first and most famous scientists to study the impact of astronomical cycles on climate was Milutin Milankovic, a Serb. In Russian literature, they like to dub him a Yugoslav scientist. Well, Milankovic did die in Yugoslavia in 1958. But he was born in the end of the nineteenth century in Austro-Hungarian Empire, thus he might as well be called an Austro-Hungarian scientist, the more so as he got his education in Vienna and worked for some time in Budapest. Milankovic got involved in a problem that stirred the minds of literally everyone and even Russian revolutionaries, like, for instance, famous anarchist, Prince Kropotkin.

The point is that, in the nineteenth century, a striking periodicity of glaciations was discovered. It turned out that once upon a time, the polar icecap was spread over practically entire Europe—the continent was covered with a thick glacial shell almost all the way to the Black Sea. And this discovery absolutely shocked the civilized world. Only two to three decades before this wonderful discovery, in Pushkin times, people knew when the world was created within one-year accuracy: one of the medieval Irish monks, while studying the Scripture, calculated the date of the creation—year 4004 before Christ's birth. In Russia, before 1700, even the calendar was based on the creation date. (But Russians were not in total agreement with the Irish monk and made an adjustment: they believed the world was created 5,508 years before Christ's birth.)

So imagine, after the slow-speed era of swords and duel pistols, the remarkable nineteenth century displayed its powers fully, the century of amazing accomplishments, the century of revolver, air balloon, Jules Verne, and steam engine, the century that turned humanity's head with successes in science and technology. Biology pointed out to humans, their ancestors—the primates. And geology pointed out to them, the world's true age, hundreds of millions of years. And finally, signs of grandiose glaciation were discovered.

Two geological features indicated the existence of a giant glacier in Europe—glacial moraines and deep parallel furrows on valley surfaces. The advancing sliding glacier works like a bulldozer with its plow—it moves earth and rock debris in front of it. As a result, a bank is created in front of the "plow" of this "bulldozer." On the land, this bank represents a series of mounds, and that's why and how they were called glacial moraines.

Besides, the advancing sliding glacier drags along some sharp rock fragments and makes long furrows with them. The glacier works like giant emery. The sharp rocks, working as cutting tools, are gradually being used up and losing their sharp facets, becoming rounded glacial boulders.

Further research by geologists showed that this kind of glacier advance is not just a one-time random event, but a periodic phenomenon.

Naturally, this discovery got all thinking citizens terribly anxious: if something like this happened in the past, could it happen again? And that could mean the end to civilization! So, the interest of Milankovic in this problem was quite understandable. Everyone was then interested in the glacial problem. Even the anarchist Kropotkin, by the way, an extremely all-round personality took a great interest in studying glacial eras.

What causes glaciation? The answer to this question is what Milankovic was looking for. It is interesting that the decision to become involved in this problem entered the head of the young teacher of the Belgrade University only after a binge. Milankovic's buddy published his collected doggerel, and his friends decided to celebrate this event in a small Belgrade cafeteria. Since teachers had little money, they celebrated the debut by sipping coffee. Suddenly, to the good fortune of not the world literature but of the world science, the cafeteria was illuminated by the advent of—no less than—the director of the Belgrade Bank, whose name was not saved for us by the ungrateful history. He inadvertently heard the discussion and got interested in the booklet. The banker liked the poems very much and bought a whole ten copies of the debutant poet's booklet. After that, the young teachers' moneys became, as you understand, oodles, and they started "celebrating" in earnest, by ordering several bottles of wine.

After the third bottle, Milankovic's buddy decided to write a big poem of all times and peoples, while Milankovic was going to "grasp the Universe and bring the beam of light to its remote corners"—at least that's how he himself later characterized his own condition in his memoirs. Eventually, Milankovic did not grasp the Universe, but instead shouldered the burden of solving the problem of glaciations.

Milankovic's idea was that glaciation is affected most crucially by the amount of solar light received by different regions of the Earth. And this in turn is linked to various parameters of Earth's rotation and revolution, like nutation, precession, and eccentricity cycle. Imagine now that the amount of light and heat, pouring onto the northern latitudes is decreasing. This means that with every spring, snow stays on the surface longer. And as you know, snow is white. That is to say, it reflects the light well. And that is to say that the planet's albedo (its reflection power) increases. And this results in a vicious circle, which, in science is called positive feedback: the less heat and light—the more and more snow, the less heat and light reach the surface, since the light is reflected by the white snow surface. Due to this positive feedback, the Earth quite quickly falls into an entirely new climate regime—the glacial period.

The glacial period lasts about one hundred thousand years, whereas interglacial periods are only ten to twelve thousand years. Our entire civilization developed during the brief period of "thaw"; we're living in the last interglacial period that is called Holocene. And by the way, this period is practically over, and if not for the lifesaving global warming . . . Well, we shall talk about it later, but now let's get back to Milankovic.

The Austro-Hungarian-Serbian-Croatian scientist, while lacking a computer, calculated amounts of solar radiation at different seasons and at different latitudes of the globe. And he did it so well that his calculated heat flows agreed later perfectly with a great number of other data already contemporary, related to times of the coming of glacial eras. Prior to Milankovic, scientists used to have heated discussions: which heat flow is more important for climate—pouring onto polar or equatorial latitudes? The summer or the winter? From Milankovic's theory, it followed that the more important was the summer solar heat flow pouring onto polar latitudes. And this was wonderfully confirmed later, too.

By the way, it is for a good reason that I mentioned earlier the plainly obvious fact that Milankovic had no computer. I did it only in order to emphasize the greatness of this person's scientific feat: it took him whole twelve years to calculate the flow of solar radiation at different latitudes of Northern Hemisphere. Yet, because of the foundation laid by him, it's clear today that it's the parameters of the heliocentric orbit that have substantial impact upon the Earth's climate at stretches of tens and hundreds of thousands of years.

But in geological scale of times, which is in the scale of tens and hundreds of millions of years, the most crucial climate factor is the ratio of land and water areas, if it's of any interest to you. Oceans are accumulators of heat. The more oceans there are, the more even is the climate and the less is the temperature contrast between the equator and the poles.

Another interesting factor influencing the climate is the average altitude of Earth's dry land. The relation here is clear: the higher, the colder. Today, the land average altitude is 849 meters (2,785 feet). Yet once it was 1.5 kilometers (almost 5,000 feet). But in this book, we are not going to discuss the issue of lithospheric plate tectonics and reasons for Earth's crust movement (although it might be included in the next one). I shall just mention that today the average altitude of Earth's surface is higher than it was in the Mesozoic period, but lower than in the Permian-Carboniferous period. The Earth is breathing, and these geological motions affect the climate as well: once upon a time, it was much warmer on Earth than it is today, but at other times, it was much colder than even in the latest glacial period . . .

Below (fig. 1) is the graph of climate fluctuations in the last four hundred thousand years.

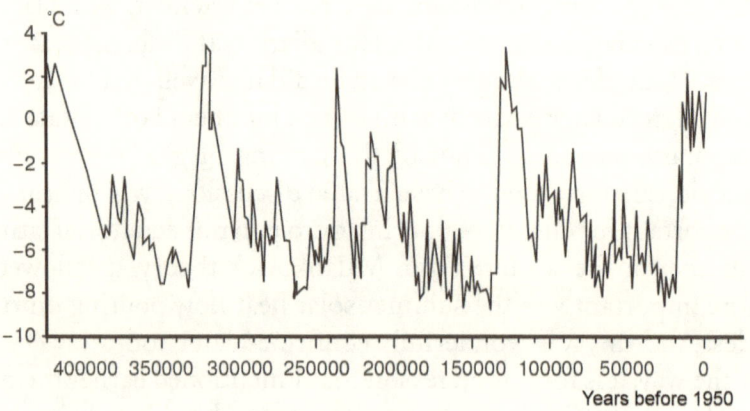

Fig. 1. Temperature changes in comparison
with the present level

The long periodical troughs on the graph are glacial periods. Brief heat bursts are interglacial periods. On the graph, we can clearly see four glacial and five interglacial periods. We are living at the very end of the fifth. Please also note how sharply the temperature drops when climate falls into glacial mode. The planet is unrecognizable in a mere period of five hundred to thousand years. The zero on the vertical axis signifies the current climate norm. (To be exact and picky, it's not current anymore, but rather it's the climate norm for the mid-twentieth century.) That is, the graph shows not the true value of the temperature, but the deviation of the average global temperature from the current one for the last four hundred thousand years. But for an obvious reason, we are more interested not in tens or hundreds of thousands of years, nor all the more in millions of years, but rather in just a few hundred and thousand years. In other words, we are interested in brief climate fluctuations comparable with the age of civilization. And I shall remind you, our civilization is about five thousand years old. What this means is that we are interested in the tiny part of the graph filling just one-half of one-scale division near zero on the time axis and in these itty-bitty jitters around zero on the temperature axis.

So what factors cause climate to fluctuate in an itty-bitty manner? And how sensitive in general is civilization to itty-bitty climate fluctuations?

We have already mentioned one factor: climate can be affected in quite an itty-bitty manner by eleven-year fluctuations of solar activity. Besides the eleven-year fluctuations, there are secular fluctuations of solar activity. Perhaps, these secular fluctuations are related to orbital periods of heavy planets. By the way, these heavy planets play a role in Earth's volcanic activity that also affects the climate.

Volcanoes . . .

CHAPTER 3

Puffing Planet

Tsar Feodor I, "the Bellringer," son of Ivan the Terrible and the last tsar in the Rurik line, was flat out gone on his head. Everybody knew that. One glance at him would enable to easily get the picture. He had a look of a very sick man with a pale, slightly swollen, half-baked face. He had a staggering gait and was always smiling, although no brick fell on his head in childhood, evidently, nature itself had a crack on him.

. . . Is it OK that I'm so politically incorrect about an imbecile?

The Polish envoy Lew Sapieha made the following conclusion after his very first seeing the tsar Feodor in action (sitting on the throne and mumbling something opaque): "Although they say he has not much brain, but I witnessed from both my own observation and words of others that he is not at all here."

It has to be noted that an old-time enemy of Russia, the Poles, took much pleasure of the fact that an oligophrenic took the Russian throne. In Krakow, they hoped that with a deal like that, things in Muscovy would come to a total decline, from which they would profit while rubbing their cool hands with joy. It surely might have happened this way if it were not for Boris Godunov.

When Ivan the Terrible passed away and his half-headed son ascended the throne, Godunov was hardly over thirty. It was this young man's shoulders to carry the heavy burden of managing the Russian State. And Godunov managed it perfectly. He was not just young and energetic, but talented as well. The ungrateful bitchy history in the face of descendants and contemporaries failed to assess this person. However, circumstances of a global scale did not give him the chance to display the full power of his talent and managerial abilities.

Godunov started well . . . He was by origin a man "from nowhere," his ancestors owing small patrimonies and playing no role in Russian history, until Boris Godunov married the daughter of Malyuta Skuratov. Thus, a new man appeared at the tsar's court, and he was soon noticed by Ivan the Terrible. By the way, Ivan's son, Feodor, "the noodle," married Boris Godunov's sister, Irina.

This Teddy (Feodor) began reigning after the death of the Terrible. But the whole brunt of managing the country fell in reality on the shoulders of his brother-in-law, Boris. It was Godunov, who established Russian patriarchy, by forcing this decision at the "international level," after some cunning intrigues—before Godunov, there were no patriarchs in Russia, and the Russian Church had foreign bosses.

Godunov successfully settled with Sweden the border dispute inherited from the "previous government." Swedes conceded to Russia the cities Russia had earlier lost: Yam, Koporye, Oreshek, Ivan-gorod, and Korela. Godunov government extended the truce with Poland and for a short while got access to the Baltic Sea.

Using a military ruse, he himself devised, Godunov utterly defeated near Moscow the army of the Crimean khan Kazy-Girey. At the time, Muscovy's situation was near hopeless with everything hanging by a thread: it was when the truce with Sweden has just expired, and all Russian effectives concentrated at the country's north. That's why Kazy-Girey was able to advance to the nearly defenseless capital and to look at Moscow from the Vorobyovy (Sparrow) Hills (5 kilometers or 3 miles southwest from the Kremlin across the Moskva River), thinking what he would do with the conquered city. This was when Godunov carried through with a brilliant operation, in which he mobilized, by way of field intelligence, Russian prisoners of Kazy-Girey. He ordered to fire cannons nonstop in Moscow. When Girey took interest in the reasons for the incomprehensible cannonade, Russian prisoners told the khan that it was Muscovites saluting from joy: fresh units came to Moscow from Novgorod and other cities. The khan got scared and turned his troops away from Moscow. Godunov skillfully used the inflicted confusion and rushed in pursuit. The Tatars were tailed after all the way to Tula, and the wounded khan hardly ducked out back to Bakhchisaray in a horse-cart.

Godunov not only fought, but also built. Building cities was his life's main work. A state is strong with its cities—Godunov understood this like no one else. On his orders, new cities were built along the Volga and its tributaries—Saransk, Perevolok, Tsaritsyn, Samara, Saratov, Tsivilsk, Urzhum, and Tsarevo-Sanchursk. The cities of Astrakhan and Smolensk were walled with new walls of stone, and in Moscow Godunov built the Belgorod wall and the Ivan the Great bell tower inside the Kremlin. (The walls built by Godunov in Smolensk and Moscow would well serve Russia in the Time of Troubles.)

In Godunov time, Archangelsk (Archangel) was founded, which immediately became an important trade center due to its felicitous location. By the way, with respect to trade, Godunov was quite knowledgeable in economics. In order to expand the trade, he allowed Brits duty-free trade, but resolutely denied their request to forbid all foreign traders other than British to trade in Russia—there must be competition!

On Godunov's orders, cities of Belgorod, Livny, Valuyki, Oskol, and Voronezh were founded in the south of the country and Kursk was under rapid construction.

Ties expanded with Georgia. To ensure access to this country, construction of a long-forgotten Russian fort was resumed at the mouth of the Terek river.

A fortified city sprang on the Yaik (currently the Ural River). In the east, things with colonizing Siberia were going on well. After the death of Yermak and departure of his militias back to the other side of the Urals, the situation with Russian colonization of Siberia seemed lost forever. And it was Godunov government, which was able to restore Russian supremacy in Siberia. Under Godunov, new cities were built in the vast Siberian space—Tyumen, Tobolsk, Tomsk, Surgut, Narym, Ketsk, Pelym, Berezov, Tara.

When a fire would occur in Moscow or some other city, Godunov would use the Treasury money or even his own sometimes to help those who lost their possessions. When an epidemic broke out in Pskov, Godunov ordered to establish sanitary and quarantine posts around the city in order to prevent the spread of disease to other areas of the country.

Another interesting psychological feature characterizing Godunov is his refusal to take the throne. On January 7, 1598, the tsar of all of Russia Feodor the Fool quietly kicked the bucket without dropping snot. The boyars then began racking their brains uphill and decided to turn Russia into a parliamentary republic, hand over all the powers to the boyar Duma. That's because the wife of the late tsar Feodor, Irina (Godunov's sister), abdicated and took the veil at Novodevichy ("New Virgins") Convent. And Feodor's brother, tsarevich Dmitry, by that time was for a long while already lying in the dank earth in the city of Uglich. And even if he were not lying there! Firstly, he was too small yet to be a tsar; and secondly, bad heredity from the crazy Ivan the Terrible affected not only the eldest, Feodor, but the youngest, tsarevich Dmitry, too. The boy suffered from epileptic fits and during one of those, he pierced his throat with a knife he was holding in his hand.

Naturally, the good old common talk decided that unlike common mortals, a tsarevich, even the one who suffers from falling sickness doesn't fall on a knife just like that. And that actually the little boy was knifed on order from Godunov. An official investigation conducted on the fact of death did not dispel the rumors, of course. Elvis cannot die! Especially so foolishly . . . And so, many a False-Dmitry appeared afterward. And some years later, when Godunov was no longer alive, the tsarevich was canonized as an innocent victim, while the common talk ascribed his death forever to Godunov. But all this was later, and meanwhile . . .

And meanwhile, the wild plan devised by boyars to put a collective tsar on the empty throne failed: the tradesmen, called to kiss the Boyar Duma cross, got agitated and started making noise and a hubbub about having never heard or known anything about that stupid Duma, and all in all gave us Boris to be our tsar! . . .

At that time, the most senior person "by rank" in the state was the patriarch like a Communist political instructor in the Soviet military, assuming command after the commander's death. So the patriarch, assuming the rights of the senior, scratched his dome and headed to Godunov to ask him to take the throne. Godunov refused. And he refused not just once or twice. Patriarch had to hold several face-to-face meetings with him, gradually moving him toward this idea.

In the end, the decision on the tsar was postponed until the Zemsky Sobor (Counties Convention), which was supposed to take place on February 17, 1598. At the Convention opening, the patriarch made a moving speech, that we don't even need to bother searching for " . . . any other ruler apart from Boris Feodorovich." The Convention supported the patriarch.

On February 20, the patriarch accompanied by a representative delegation of clergy and people arrived at Novodevichy Convent where Godunov was staying with his sister Irina. The patriarch and the entire delegation tearfully and humbly pleaded with Godunov for his taking the unoccupied throne, but once again, they received a refusal.

The next day, an impressive demonstration of citizens moved to the Novodevichy Convent, they were marching with icons, gonfalons, wives, and small children. If it were Nick the Second in Godunov's shoes, he would have created a Bloody Sunday by shooting two or three volleys. But Godunov was a soft person. So he agreed to become the tsar.

The start of Boris Godunov's reign was highly promising. He relieved all peasants of duties for one year, merchants were relieved of any duties for two years, and heterodox believers got full tax relief for one year. All financiers got a whole annual salary as bonus. Amnesty was declared. The poor and those with no social protection (widows, orphans, paupers,) received monetary allowance. Godunov abolished death penalty.

During Godunov's reign, a perfect job was done by the then-Russian analogue of FEMA—if a flood or fire happened somewhere, food products, warm clothing, and money were immediately sent there. Godunov, like Putin, traveled a lot about the country. There were no fighter planes yet at that time, so the sovereign traveled by cartage, attempting to be the first at the place of emergency with humanitarian goods. Never before did Russia have such an active tsar.

Godunov decided to establish secondary schools in Russia and build a university. The idea that it was time for Russia to join Europe in culture development stuck in his head like a nail. It was Godunov to become the first of all the Russian tsars to send talented youth to study abroad. He used all means available to try to bring educated foreigners and scientists into Russia.

It might seem that with a ruler like that, it was time to live and prosper! It might seem that his bright name should remain in history forever and eclipse with his renowned and humane deeds the atrocities of Peter I, "the Great." And properly speaking, there would be nothing left for Peter to do after Godunov, but . . .

But for some reason, instead of soaring, Russia after Godunov fell to the abyss of Time of Troubles. And for some reason, one of historians writes about Godunov the following sad and moving words: "A brilliant victory was evident, but such was Boris's ill-fated fortune when each of his successes turned in public conscience into something harmful to him. Not a single statesman in Russia before him would spend so much efforts in order to deserve people's favorable disposition ... Godunov's name was never written in execution decrees, but it was always written in all those bestowal or award papers, yet in spite of this ... there is no such sin that he would not be accused of, while often totally without grounds or proof. For instance, after the flight of Kazy-Girey, rumors started to circulate that Boris himself brought in the khan to divert people's attention from Dmitry's killing."

And I should add that after the Moscow fire, rumors were circulating that the capital was intentionally set on fire on Boris's orders to give the tsar an opportunity to help the victims who lost their possessions and to earn thus cheap popularity. Following Nero's footsteps ... And you still wonder why I don't like a people ...

So what has caused the dramatic change of Russia's history at its most promising time?

Volcano

Before discussing volcanoes, a few words about the climate of that era. After the twelfth century AD, the Earth rolled down into a climate regime dubbed by historians and climatologists the Small Ice Age, which lasted "with variable success" until the late nineteenth century. It must be said that the name devised by scientists was quite well turned. In fact, many surprising things happened during this micro-glaciation: in Venice, they were ice-skating; in Cologne, it was snowing on June 30, 1318. And in Russia between the fourteenth and nineteenth centuries, snow could fall generally in any summer month. Occasionally, the summer frosts recurred several years in a row. It happened in the fourteenth, fifteenth, and in the late sixteenth centuries and caused, of course, bad harvests and famine.

But the early years of the seventeenth century were hit the hardest. It was the start of Boris Godunov's reign. The lean years came in a cluster, as scientists would say, or one after another in 1600, 1601, 1602, and 1603, it snowed in Moscow in both July and August, and frosts happened regularly in all summer months. What kind of harvest could be expected in such conditions? It's possible to keep on going through one lean year and use up supplies from the previous year, but it would be practically impossible to survive two and all the more three years with no harvest. And here they had four lean years in a row!

Contemporaries noted that "neither grandfathers, nor great-grandfathers" could recall times like these. The famine was such that people ate up all the dogs, cats, mice, and rats. They were eating even their own children. At farmers' markets, they were selling human meat. It was dangerous for a traveler to spend a night at a coaching inn, for the risk of simply being knifed and eaten.

Naturally, Boris tried to fight the nature as hard as he could. He ordered to unseal strategic stocks of grain and distribute them free of charge to the starving; he dispatched the grain to those provinces that suffered the worst from the bad harvest; he gave money to the poor so they could buy food; he tried to freeze the prices of bread. Since the money budgeted to fight the famine was mercilessly cleaned out by the bureaucrats, Godunov turned to direct food supplies to provinces.

All the dead were buried at the expense of the Treasury. And by no means was it cheap for the Treasury: in Moscow alone, over one hundred thousand corpses had to be buried. The country was overwhelmed by gangsterism and religious fanaticism. Gangs of robbers stopped everyone and especially on main roads, thus breaking up communications. This reached the level when a huge gang of bandits under the command of the gang-leader Kosolap (*the name translates as a nickname for Bear—Translator's note*) appeared at the walls of Moscow. Against this gang, Godunov put forward the regular army that was able to get the better of the gangsters, but only with great difficulties. That was the first wave of Big Troubles . . .

People—What ungrateful animals. Who is to blame for the bad harvest? A strange question. Who is always to blame for everything? The tsar, of course! He is responsible for everything in the country! Evidently, Boris displeased the Lord—otherwise why would the Almighty punish the Russian land so hardly. Evidently, the God is avenging the death of tsarevich Dmitry, the Moscow fire, the Godunov's bringing to Moscow of Khan Girey, and that foreign scientist-infidels were invited to the Holy Russia. The Lord takes revenge on the tsar, but the people suffer! What a scum, this Godunov! And do you know why he gives bread to the starving? To prey for his sins, swine! And by the way, there's not enough of tsar's bread for everyone! Hence, the tsar must be stealing the people's bread.

All sorts of crazy rumors always spread about all the rulers of all the countries. But all the gossip and the most stupid cock-and-bull stories about Godunov were scrupulously collected by history and preserved in the people's memory. Along with the nickname of Godunov's reign—Miserable—Miserable Reign . . .

But just wait, just wait! The Savior cometh! And who knows, maybe, little Dmitry, killed for no reason, will rise from his coffin and save the Holy Russia from the foe! Revenge will come to the wolf for sheep's tears!

Elvis Is Alive . . .

The rumors that the epileptic who fell onto the knife and cut his carotid artery, miraculously survived, were spreading for a long time in the depths of popular masses, but it took a natural catastrophe to materialize this rumor. And many a False-Dmitry gemmated.

Being a reasonable person, Godunov understood the threats to the state from this idea of a resurrected Dmitry. The living tsarevich was less scary to Godunov than the dead one. And it seems Boris, in his time, took the idiotic death of Dmitry no less hard than Napoleon took the tragic shooting of Duke d'Enghien. They both sensed that the deceased would not forgive them those accidents . . .

On October 16, 1604, the miraculously resurrected tsarevich Dmitry with a big wart on his nose (the one inherited from Greg Otrepiev) and a small detachment of Polish interventionists entered the bounds of the State of Muscovy. Cities were surrendering to the Savior one after another. Service class deserted to the wonderful tsarevich in huge flocks.

Godunov tried to hold out as hard as he could. False-Dmitry, besieging the city of Novgorod Seversky, was thrown back by the troops of the legal sovereign. Then the troops of the Moscow tsar under the command of Feodor Mstislavsky gained another victory over False-Dmitry and forced him back to Putivl. On his hand, the political instructor of Moscow and of whole of Russia, Patriarch Jove made a resounding statement that the son of Ivan the Terrible, Dmitry, is not alive and the person calling himself Dmitry is a fugitive unfrocked monk Greg (Gregory) Otrepiev. The statement was read in all churches of the country. But people no longer took anything into the head, but "chose by heart," thronging in flocks into the impostor's army.

. . . I have mentioned earlier that people—they're ungrateful pigs! Nevertheless, I repeat . . .

It's hard to say what the country's fate might have been if Godunov's inner circle were not to waver and to poison him. The tsar woke up cheerful and jolly in the morning of April 13. But having finished his lunch, he soon started complaining about feeling sick and stomach pains. He started bleeding from nose and ears and at about three o'clock in the afternoon, the tsar passed away—"from great grief," as they would say later. And coming next, is the Time of Troubles—the Time of Timelessness.

And now perhaps is the moment to reveal to the reader as to what has caused such a catastrophic drop of the annual average temperature. I already mentioned that Godunov's reign took place at the very middle of the Small Ice Age. And in accordance with the theory supported in this book, worsening of local climate leads to soaring of human spirit, to new and unusual inventions, and to the birth of great empires.

So, that's how it was going on in Russia. Ivan the Terrible was building the empire. Godunov replaced him and successfully continued what had been started. But then it turned not just cold, but catastrophically cold. Climate made a leap down so sharply and strongly that the social system did not have enough time or plain ability to adequately react to it. It collapsed. "Every significant relationship has extreme nature"—those who read my book *Monkey Upgrade* know this great Nikonov law. If it gets cold—it's good, of course, for building an empire. But if it gets excessively cold . . . Then things happen the way they happened with Godunov, Russia . . .

Scientists knew since long ago that summer frosts and clusters are usually related to large eruptions of volcanoes throwing into the atmosphere so much junk of all sorts that atmosphere's translucence is reduced causing a "nuclear winter" that lasts several years. In first approximation, this is the actual mechanism of the temperature drop (corrective technical details to follow). In general, all signs indicate that the cold cluster of 1600-1603 was caused by a large volcano eruption, but which volcano did erupt?

Only relatively, recently, the location was finally determined of the ill-fated eruption that caused a national catastrophe for Russia. It turned out that in 1600 in South America, in the territory of Peru, Huaynaputina volcano erupted. The culprit was found by an American glaciological expedition that was working on glaciers in tropical Andes. A poor Russian scientist Mikhalenko drilled the ice at a giddy altitude of 5,000 meters (16,000 feet), and Americans then studied the chemical composition of the ice against the years. It is no more difficult to count the annual rings in ice than it is to count them on tree-cuts. In the layer corresponding to the year 1600, elevated amount of sulfur was detected, directly pointing to a powerful eruption.

Volcanoes, these unpleasant and dangerous boils of the planet, are torturing us for as long as we exist. Thousands of volcanoes are on the planet (nobody knows their exact number) and not all of them, by the way, look like boils. There is so-called fault volcanism—it is rather the sores on the planet's body. An ordinary person would not even call them volcanoes because they lack the usual cone-type shape. These "under-volcanoes" are in Iceland, Canary Islands, and Hawaii. They look like this: lava with small amount of gas content pours from cracks in earth's crust slowly and constantly. Hawaiian volcano Kilauea, for example, erupts daily during many centuries. This phenomenon even got its proper name—Hawaiian type of volcanism.

. . . Oh, if only all eruptions would proceed so intelligently! Then Briullov would not have painted his famous painting. And the world art would have suffered a significant loss . . .

Volcanoes are spread very unevenly over our planet. For example, Europe practically lacks them (except for the one single Vesuvius on the mainland and some volcanoes on islands in the Mediterranean and Iceland). There are no volcanoes in most of Siberia and continental Australia. At the same time, 80 percent of all eruptions are produced by volcanoes of a so-called Pacific Fire Belt. The wittiest readers immediately guessed from this name that the Pacific Ocean is encircled with a volcano belt. Indeed, recall the map and move counterclockwise—the Aleutians, Kamchatka, the Kurils, Japan, Taiwan, the Philippines, Indonesia, the Solomons, the Hebrides, New Zealand, then a volcano chain stretches along the Pacific Ocean bottom, after that, volcanoes are seeded along the entire western coast of both Americas, and finally we're back at the Aleutian Islands. The belt.

This belt represents the boundaries of the huge Pacific plate. The most powerful eruptions always happen in the zones of subduction (collision of lithospheric plates), and the enormous energy that splashes out in form of eruptions is none other than the energy of friction of one lithospheric plate against another. Volcanologists observe volcanoes thoroughly, catalogue them, and publish monthly bulletins with descriptions of what erupted and with what force. These are very important observations, but unfortunately, they are useless: they have no forecast value. Alas, volcano eruptions, as well as earthquakes, cannot be forecasted as of yet.

Here is just one illustration to this thesis. In June 1993, international volcanology conference took place in Columbia. In the course of it, the scientists attending the symposium decided to visit the well-known Columbian volcano Galeras. The tour matched the topic, as they say. And a strange coincidence happened: just as the group of twelve volcanologists ascended to the crater, eruption started. Not a very strong one, it was just three points. However, nearly all the "tourists" perished on June 7, 1993, the cream of the world volcanology has fallen on the slopes of Galeras. And it's not at all funny. It's a tragedy, dear readers. We still know too little about volcanoes. However, we do know something.

The most dangerous are the volcanoes of the so-called Pliny type. They got their name from Pliny the Elder, who was the first to describe this type of earth pimples. Pliny the Elder was a citizen of Pompeii. He described one of the eruptions. But during the other one, on August 24, 79 AD, he died. He was a very consistent person . . . Pliny volcanoes are the dormant volcanoes. They could also be called explosive volcanoes. The rocks that constitute Pliny volcanoes harden and plug the crater just as a cork plugs a bottle of champagne. And then, when unexpectedly to everyone around, the cork is ejected, the troubles begin.

Unpredictability of volcanoes often surprises even volcanologists; for instance, the major eruption of the twentieth century was of a volcano that was not even in the list of active ones. It happened on June 15, 1991, on the largest Philippine island, Luzon, just 80 kilometers (50 miles) off the Philippines' capital. Mount Pinatubo exploded. The number of people who died in the eruption was eight hundred, and over hundred thousand became homeless. The relatively mild consequences were only because this eruption wasn't the strongest in Earth history.

Pinatubo exploded, throwing out ashes and gases up to 24 kilometers (15 miles) into the air. Clouds of dust and ashes reached Singapore that is 2,400 kilometers (1,500 miles) from the volcano. Flows of melted rocks heated up to 1000°C (1800°F) gushed down the slopes in rivers of fire with speeds of a high-speed train. Up to hundred kilometers (miles) around, day turned into night. In the province suffered the most, Zambales, everything was covered with a one-meter (three-foot) layer of ashes and volcanic debris. During the eruption, about thirty million tons of sulfur dioxide was thrown out into the atmosphere. The tiniest particles of ashes formed a huge cloud that encircled the entire globe at the equator.

But an eruption far more powerful occurred on August 26, 1883. When I name the volcano, you will probably recall something—Krakatoa. This volcano is located on the island by the same name in the Sunda Strait between Java and Sumatra. When it exploded, the plume of hot gases reached altitudes of 80 kilometers (50 miles). The rumbles were heard as far as 1,000 kilometers (over 620 miles). Nineteen cubic kilometers (4.5 cubic miles) of all sorts of junk was ejected from the crater, and the ashes were falling out in the radius of over 5,000 kilometers (3,000 miles) from the island. The eruption caused a tsunami 30 meters (100 feet) high; it encircled the globe and its effects were felt as far as the English Channel. For comparison, the recent deadly tsunami in Southeast Asia, still memorable to many, had waves of only six meters (20 feet) high . . .

Krakatoa destroyed about three hundred cities and towns and took away the lives of thirty-six thousand people. The press was already in existence then, as was the telegraph, and that's why the Java catastrophe received wide social resonance. Unlike the much more powerful one, Sumbawa. However, before discussing Sumbawa, a few words about how eruption power is measured. It is measured in points—from 1 to 8. But it would be wrong to think that a five-point eruption is only one quarter stronger than the one that received four points from scientists. Oh no—it's ten times stronger! Every point represents a tenfold increase in eruption power.

Well, during the entire time of systematic observations of volcanoes (for about 100 to 150 years), there was only one eruption that received a seven—it was the Tambora eruption.
On April 5, 1815, on Sumbawa Island in Indonesia, Tambora volcano exploded. The rumbles could be heard at a distance of 4,000 kilometers (2,500 miles). For several days after the eruption, it was pitch dark over Sunda Archipelago—the sun was totally covered by volcanic dust. The volume of material ejected from Tambora was 400 km³ (100 cubic miles), which is half of the Lake Ladoga volume—the largest lake in Europe. But considering that volcanologists measure ejected material in so-called dry rock equivalent, which is reduced to density of 2.5 g/cm³ (1.5 oz./in.³),

while in fact volcano ejects foamed rock masses, then the real amount of Tambora ejections could fill the Ladoga basin to the top and even above. Following this natural catastrophe, the average altitude of Sumbawa Island lowered from 4,100 meters (13,500 feet) to 850 meters (2,800 feet) above sea level. So the small island was literally blown apart by the explosion. That was some "seven" . . .

However, our planet's history witnessed even stronger eruptions—the ones deserving a whole nine from volcanologists if they were to be around then! And each such eruption was accompanied by historic events such as migration of people. It was not the eruption itself, of course, that caused mass migrations, but the cooling that followed.

It has to be said that although there were never any volcanoes on the territory of Muscovy, its history includes a number of dark pages written by volcanoes. For instance, analysis of ice samples shows that in 1258 somewhere on the planet (although it's unclear where) a strong eruption rumbled—according to the peak on the graph, the blast was about two or three times as powerful as Tambora! Consequence: the chronicles for 1259 from Novgorod and Pskov noted an unusually cold spring on May 31 (practically summer!), strong frosts hit, while the autumn came so early that the entire harvest was buried under snow.

By the way, what is the cause for global cooling? Having read up to here, you might consider the question idle, perhaps. Indeed, if clouds cover the sky, the air smells hell knows what, the nose is full of dried dust flicks, and the sun isn't able to get through the darkness that fell over the city hated by the procurator, then why ask strange questions? The answer is clear: when the sun doesn't shine—the earth can't warm up. However, my question is not as strange as it may seem.

Take a banknote, a bill that is most popular in Russia—a one-hundred dollar bill, you surely have one. Take a good look at the face of the person portrayed on it. It's Benjamin Franklin. Not the American president, as many think erroneously, but a great scientist, one of the authors of Declaration of Independence, and in the 1780s, the US ambassador to France. The latter circumstance is important in context of this book. Because it was at that particular time, when the inquisitive American resided in Europe, that the Laki volcano in Iceland erupted, leaving behind almost 100-kilometer-long (62 miles!) river of hardened lava and dry fog that blanketed entire Europe in the fall of 1783. After that, an unusually cold winter came.

This chain of circumstances—eruption-fog-cold—did not go past Franklin unnoticed and led him to an idea that dust affects climate. For that time, it was a good hypothesis. In fact, it was so good that it stood for almost two hundred years, up to the end of the twentieth century. And only then, when studies of the upper layers of the atmosphere began with the use of flying vehicles and when new methods for measuring the transparency of the atmosphere appeared, it was revealed that it was not the volcanic dust, which was responsible for global cooling. The dust hypothesis was replaced with the aerosol one.

The dust falls out quite quickly. Besides, it's able not only to cool, but also to warm up the atmosphere, and it all depends on the size of dust particles: small ones (about one micron) cool the atmosphere while larger ones warm it up. As to the aerosol . . .

What is the essence of the aerosol hypothesis? Let's look once again at Tambora. It ejected seventy million tons of sulfur in the form of sulfuric gases. The gases were hydrated by the atmospheric moisture and created sulfuric acid that formed a fog. Unlike the dust, this fog does not fall out quickly but rather prefers to stay in the upper layers of the atmosphere for years, where it reflects sunbeams and thus causes climate mess.

The structure of the atmosphere is such that if a volcano erupts in high latitudes, the sulfuric-acid volcanic aerosol blankets only that hemisphere of the planet where the eruption occurred. In other words, if an eruption were in the Northern Hemisphere, the cooling would occur in the Northern Hemisphere; if eruption were in the Southern Hemisphere, cooling would be in the Southern Hemisphere. And if a volcanic explosion happened in lower latitudes, not farther than 15 latitudinal degrees from the equator, then the aerosol is spread over both hemispheres and the cooling occurs worldwide.

The Pinatubo explosion lowered the average global temperature by 0.3 degree Celsius (0.5°F). Is it much or little? In the twentieth century, the earth warmed up by 0.6 degree (1.1°F), causing everyone to shout about global warming. That is, one relatively small eruption immediately ate up half of the global warming! And what if there would be a big eruption? True, the aerosol from Pinatubo dissipated in three years, but it is known that aerosol from big eruptions can be present in the atmosphere for decades. Then it would be time for turning on the positive feedback with polar ice albedo, which we have discussed earlier, and the globe would quickly fall into a real ice age. Not small, but big one. In other words, one extremely super-large eruption or a series of large eruptions could serve as a trigger mechanism for this terrible process.

A reasonable question: are large eruptions expected in the near future? A discomforting answer: they are expected. For the simple reason that it's been too long a time without them. Besides, the elevated volcanic activity correlates well with orbiting cycles of heavy planets (we mentioned this earlier) and with periods of inactive Sun of the Maunder type (this too was discussed earlier). Our future awaits us with a period of inactive Sun and thus elevated volcanism.

The twentieth century was a century of relatively calm volcanism. It had several unimpressive eruptions and a couple of more impressive ones. In 1991, Pinatubo exploded. On May 8, 1902, a volcano erupted on Martinique, resulting in total destruction of a flourishing city of St. Pierre. All of its thirty thousand inhabitants perished. Only one person miraculously survived, he was in the underground prison. He got severe burns, and for the rest of his life his only occupation was to demonstrate his scars when traveling with a vagrant circus in America. Today there is an eruption museum on Martinique. Those who visit it say that the museum leaves about the same impression as the Hiroshima museum—the same panorama: a city where not a single building remained. Gray desert. Moon landscape.

Once again, due to the mass media and telegraph, this eruption received wide publicity. But according to the scale mentioned earlier, it hardly measured a four. Tambora was a thousand times more powerful and its performance received a grade of seven points.

But could something like that happen not in a land of Papuans but in a high-populated area? It cannot be ruled out. Moreover, this prospect is quite real. In Germany and France, for instance, so-called lava fields are located, they are remnants of Quaternary volcanism. These are long extinct volcanoes. Once upon a time, eruptions happened here with power comparable to the Tambora eruption. And the fact that these volcanoes are long extinct should not fool anybody. As the phrase goes, all new is well-forgotten old.

And was our planet severely shaken long ago! A real beauty . . . If you are ever in Singapore, don't be lazy to fly to a big city of Medan on Sumatra Island. It's a one-hour flight, and after a two-hour car ride along a nice road, you are at the Toba Lake. The lake is not small from one of its coasts you can hardly see the opposite one. This is the crater of the Toba volcano, which exploded seventy-three thousand years ago. Volcanologists ought to mark Toba with the full eight. That is, the power of its explosion was ten thousand times greater than the power of the volcano that razed to the ground the city of St. Pierre.

The volume of the ejected material was 4,000 km³ (1,000 cubic miles) in dry rock equivalent. The mathematical model of the explosion gives the scale of surface cooling in the Northern Hemisphere of 3.5°C (6.3°F) in the course of ten years. The very real ice age! And it has indeed come. It was the time when the planet was kind a fluctuating in a transition era from interglacial to glacial, and the weakening climate system got a fatal blow. When finally the volcanic aerosol dissipated, it was already too late: the power of glaciers was such that the climate system could not recover and the glacial period lasted another sixty thousand years.

. . . That's what we should expect and dread . . .

By the way, according to one of the versions, the reason for the current global warming is the volcanic calm of the twentieth century. According to the opinion of some climatologists, this calm is the main thing constituting the basis of climate prosperity that we are experiencing today.

After the events of 1902, when a volcano killed the city of St. Pierre, in 1912 another volcano, Katmai in Alaska, scared the aboriginals. And then for almost half century came the calm. There was no such calm for fifteen hundred years! Isn't this perhaps the calm before the storm?

According to forecasts, the intensity of volcanic activity should start increasing as soon as in the twenty-first century and could equal the activity of the early nineteenth century. And the nineteenth century was a cold one, if you remember—it was not without reason that Dostoyevsky's characters were wearing overcoats in summer in St. Petersburg.

CHAPTER 4

Gases!

So, what have we learned? That climate depends on activity of the Sun, on position of Earth on the orbit, and on volcanic activity. Those with advanced brains, who heard with half an ear something about Kyoto Protocol, could also remember something about greenhouse gases.

Yes, greenhouse gases . . . As famous Soviet comedian Raikin said, "Everyone inhales oxygen, but aims to exhale all kinds of muck!" It's about this muck that we're now going to say a few nice words.

The Earth's atmosphere is nitrogen and oxygen. It has 78 percent nitrogen and 21 percent oxygen. This adds up to 99 percent. And the remaining 1 percent covers all the rest—argon, carbonic acid gas (carbon dioxide), methane, hydrogen, etc . . . It's this 1 percent being so crucial for the climate, as strange as it may seem. The most important part is played here by three—and multi-atom gases, such as CO_2 or CH_4. Two-atom gases (O_2 or N_2) are transparent to short-wave radiation from the sun and to long-wave radiation reflected from the earth's surface. But three-atom gases are only transparent to those radiation beams that come from the sun. While the beams that are reradiated (reflected) by the earth are stopped.

. . . Now wait a minute, is it OK that I wrote those chemical formulas, uh? Are you overstrained? I'm setting my hopes on clever and intelligent readers who did not shirk school, especially chemistry, and know that CH_4 is methane and CO_2 is carbon dioxide that was exhaled by above-mentioned Raikin. Oh Lord, do send me such mighty intellectuals . . .

Thus, the more amount of three—and multi-atom gases are present in the atmosphere, the more heat is stopped by the atmosphere, since these gases do not let the reflected radiation pass through, but rather absorb it, causing the atmosphere to warm up. It's similar to a greenhouse effect, which is the reason for calling these gases "greenhouse" gases.

In the last 150 years, due to human activities, concentration of the main greenhouse gas—CO_2 in the atmosphere increased by more than one third. And it reached prodigious figures—0.038 percent. For comparison, argon amounts to 0.93 percent in the atmosphere, or almost twenty-five times more. However, all that oodles of argon has no effect on climate, but hundredth fractions of carbon-acid percent certainly have. It's a subtle thing—climate . . .

Carbon-acid gas, known better as carbon dioxide, is exhaled by every living thing, it's produced with high efficiency by industry, it's hated by Greenpeace members, its discharge is under attack by liberal democratic politicians set up by the "greens" aiming at its reduction, and we shall definitely discuss it later when we'll be vivisecting the Kyoto Protocol. But now, our song will have a verse about methane.

Methane is discharged by bogs. That's why the total area of bogs is certainly taken into consideration by formulas in climate mathematical models. Methane is also discharged by sheep and generally by all large mammals. Sheep fart with methane. That's why the total number of sheep, goats, and cows, amounting, by the way, to hundreds of millions head in the whole world, also enters in the climate models.

No doubt, humans contributed their mite for saturating the atmosphere with carbonic acid. Moreover, calculations show that it was human industrial activities, which moved back the start of the next ice age. But this is yet another topic that we shall discuss in due time and in due place . . .

And now, let's talk about another unexpected factor that affects the climate. It turns out that climate is affected by . . . climate itself. Or to be more exact, by the complexity of the climate system. Modeling on powerful computers visually demonstrates this. When the planetary climate model is mathematically reproduced and run on a computer, without being loaded with any external exposures, that is with volcanism turned off, with content of greenhouse gases in the atmosphere stabilized, and with fluctuations of the Earth's axis and solar activity excluded, then the system still experiences strong fluctuations even in the absence of all external perturbations. Monsoons and trade winds are blowing, precipitation amount is changing, and the average global temperature is floating by 0.4-0.5°C (0.7-0.9°F).

Climate turned out to be quite a complex system to be able to live a life of its own even without external perturbations. Systems of such complexity create random events themselves and process them themselves. It's established today by science that the climate system has its inherent very stable cycles—one hundred thousand years, forty-one thousand years, twenty-three thousand years, 2.5 thousand years, two hundred years, 65, 22, 10-11, and 7.5 years . . . They are superimposed over one another, producing a very complex picture. Because of this complexity, the weather picture of the planet seems at first glance a total chaos. But this seemingly total chaos has its own order.

CHAPTER 5

UNESCO and a Child

In 1972, a Soviet oceanologist, Andrey Polosin, Ph.D., was working as head of physical oceanography department of a scientific institution. And it was he to receive a request from quite a non-Soviet UNESCO Oceanography Committee to write an analytical paper about El Niño. Apparently, there was no better specialist in the world at that time than Polosin. But why UNESCO all of sudden got so interested in the El Niño problem, that it requested a paper by a scientist from the communist empire?

El Niño is Spanish for "a boy," or "a male child." That is the name used in South America for a mysterious periodic phenomenon, which used to bring troubles only to the western coast of the continent and which in 1972 for the first time in human history became the cause of the world's economic crisis. Today many have heard about El Niño. The most literate philistines define El Niño as an "irregular current." We are not going to argue with dabblers—after all, they still tell us that the Gulf Stream is a warm current pictured as a river of warm water flowing in the ocean. But neither the Gulf Stream is a river, nor even less so is El Niño. The Gulf Stream is, let's say, a front of interaction between warm tropical and cold Arctic waters. But oceanology is not the subject of this book, so let's get back to climate and move to the western coast of South America.

We're taking off! We're flying over western South America in the area of Peru, where rocky coast steeply goes down into the ocean abyss. What a beauty! Hot sun. Blue sky. But it's not hot. Because from the land and toward the ocean, trade winds are blowing. Trade winds in general are considered the most stable winds in the world. They blow at the equatorial area from east to west day and night in summer and winter. But their force varies. Occasionally in summer—usually, late December or early January (keep in mind that in the Southern Hemisphere summer is when we have winter)—trade winds become so weak that they practically don't blow at all. That's when El Niño appears. This phenomenon is named after Christ the boy, because it happens usually after Christmas. However, it's not at all as innocent as was the newborn Jesus. Rather the opposite.

Under normal functioning mode, the climate mechanism works like this: constantly blowing trade winds push the waters heated by the hot sun away from the South American coast into the ocean, and they are replaced by cold waters from the ocean depths—this phenomenon of lifting deep waters is called upwelling. Besides, from Tierra del Fuego (Land of Fire), north all along the South American coast, the cold Humboldt or Peruvian current flows. It starts at Antarctic latitudes, flows north almost to the equator, and only at the latitude of Ecuador, Cape Pariñas projecting far into the ocean thrusts this cold current west, away from the continent. That's why the only place where swimming is possible is in Ecuador, north of this cape. While south, it's chilly. This is all rather strange, of course: the equator is near, the heat is almost unbearable, but it's impossible to get into the water! The water temperature is about the same as in the Black Sea in late fall; the water is so cold that penguins live along the entire South American coast. Just imagine, penguins at the equator!

It wasn't without reason that the great Andean civilizations were not the civilizations of seafarers—neither Incans, nor Aztecs built ships. Not only did they never sail into the open ocean, but even didn't go along the coast. And it was not just because the water was cold, of course, but also because due to all these strange weather things, the swash over there is crazy, it's called dead swell. Even in the calmest weather, huge waves roll on the shores, threatening and able to crush any wooden and even more so a reed longboat. Representatives of the ancient American civilizations bravely sailed only in the Lake Titicaca, it was calm and safe there. Europeans were much luckier in this sense that their seafaring alma mater was the quiet, warm, and calm Mediterranean Sea (for details about this, see my book *Civilizer's Fate*).

At this place, perhaps, someone would recall Thor Heyerdahl who, in order to prove that Easter Island was populated by migrants from South America, upped and sailed on a reed raft from South American coast to Easter Island. Kinda, they were seafarers, the Indians, since they got to Easter Island! But I don't advise taking amusing experiments of Thor Heyerdahl seriously. It's not without reason that many people justly believe that he was a man who spent his entire life for the one and only purpose of self-PR. Indeed, it's possible to sail on a raft to Easter Island. Perhaps, with some good fortune, it would be even possible to do it in a big pan. However, humans in general were populating the planet not in the way Heyerdahl imagined. First humans got to Easter Island from Central Polynesia (fig. 2). And they got to Polynesia from Melanesia and New Guinea. And they got there from Asia thirty-five to forty thousand years ago, in the great glacial era, when the ocean level was much lower and New Guinea and Sunda Archipelago were one and had land connection to Asia. Also, Australia could be reached by land! But New Zealand was for a long time completely uninhabited for the reason that it never had a land bridge to major areas of land . . . What else is left there? America? Humans came to America by way of Bering Isthmus at approximately the same time they came to Australia.

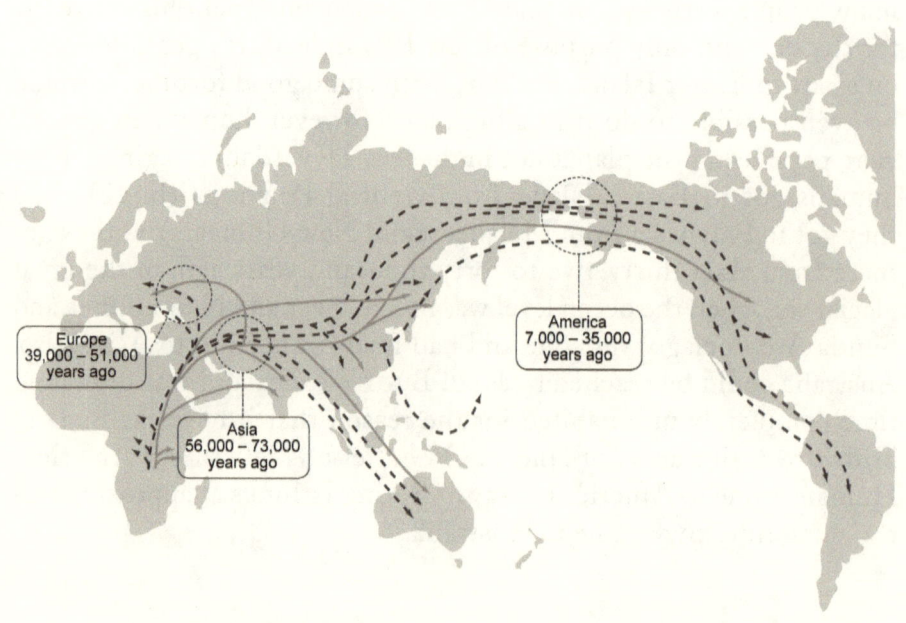

Fig. 2. Paths and dates of people's settlement by genetic data: determined by mitochondrial DNA types; determined by Y-chromosome types

Well, OK, let's get back to Andrey Polosin. Right now, he is sitting at his office desk filled with foreign and Russian scientific magazines and writing a report for UNESCO. Let's take a peek over his shoulder.

The cold water, brought by the Humboldt current from Antarctica, and especially the water that rises from the ocean depths, is very biogenic, or biologically active, because it's enriched by beneficial microelements that greatly promote plankton development. Plankton is the food for a small fish anchovy. Anchovy is another name for sprat and other mini fish. But it was fishing for them that made Peru the world's greatest fishing power. As great as the USSR and Japan. Both the USSR and Japan caught about ten million tons of fish per year. And the same amount of ten million tons was caught by Peru. But while the USSR and Japan used huge seiners and trawlers to catch fish all around the world oceans, Peru used small scows and longboats. And they fished not all around the world oceans but in a 100-km (60-mile) coastal strip. It's abundant, the anchovy fish! . . . The cold Peruvian water is so amiable to anchovy that 80 percent of its world's catch happens here, in Peru. Anchovy in turn serves as food for larger fish. These larger fish are food for birds and other critters—sharks, dolphins . . . It's indeed rich and diverse, the animal kingdom of the western coast of South America!

But sometimes the wonderful wind-water mechanism, ensuring prosperity to anchovy, breaks. It happens at those specific rare cases when trade winds calm down their blowing just before Christmas. And since there is nothing that could thrust away the warm waters from coastal areas, it stays there in the form of a ten-meter (thirty-three-foot) thick layer like a film of oil, blocking the oxygen from accessing the depths (we remember that the warmer the water, the lower is the solubility of gases in it). Under the equatorial sun, the water temperature quickly reaches 30°C (86°F), and life in the coastal strip comes to a standstill—suffocates and "boils to rags." Plants and dead fish are decomposing, thus actively emitting hydrogen sulfide. Millions of tons of dead matter produce so much of this stinky hydrogen sulfide, that upon its oxidation in air it leaves ugly black stains on the sides and sails of passing ships. It creates quite an unpleasant if not a grave impression. This phenomenon even got the name "Callao Painter."

Small-scale El Niño happens once every two to three years, when Christmas trade winds moderate their temper, and it goes unnoticed. But once every ten to twelve years and occasionally more often, trade winds lose their strength so much that a catastrophic El Niño occurs. Super El Niño!

Tens of thousands of tons of dead fish, crawfish, jellyfish are thrown out onto the shores, where all this rots under the tropical sun, ponging unbearably. In addition to tons of sea critters, the pong is spread from tons of dead birds killed by lack of food. And the Peruvian army in gas masks uses heavy equipment and bulldozers to bury rotting meat into ditches to avoid epidemics . . . By the way, sea birds produce the world's best organic fertilizer, the famous Peruvian guano, the stuff that Peru is proud of and exports. No anchovy—no birds—no guano—no export. But the losses from underexport of dung are not, of course, the main trouble. The worst problem is that anchovy dies out! For it's the anchovy that the economy of Peru is largely held up by.

Exactly by that same unfortunate year when the catastrophic El Niño happened, Peru reached the mark of ten million tons of fish-catch. Of course, nobody could be able to eat that much anchovy. That's why more than 90 percent of anchovy was processed into chicken fodder. All European poultry farms were buying cheap fodder made of dried and ground anchovy.

And all of a sudden, there was no fodder. Birds were dying. Prices on poultry meat skyrocketed. This led to price increase on beef, pork, and mutton. In the unfortunate South America, where meat became prohibitively expensive for the population, people started eating beans. So, naturally, prices on beans also skyrocketed. And this was just aggravated by the situation in agriculture (lack of fertilizer—guano).

In old days, El Niño, even a catastrophic one, was a problem only for South America. It was not even a problem, but rather a small local unpleasantness. It stinks . . . But nowadays, in the times of gradually globalizing economy, things are completely different. It's known that the Pacific Ocean produces twenty million tons of anchovy a year. When Peru was catching one million tons, there was enough fish for all—birds, humans, and even death-doing Moloch El Niño. But when humans started taking out a half of what the ocean produced, even the smallest malfunction was enough to cause a catastrophe. And it occurred. World markets became unsettled. Political situation in South American countries became unstable. Even the Central Committee of the Communist Party of the Soviet Union (CPSU) became concerned, they never before could imagine that the world was so small and intertwined.

—

Why did I tell you about El Niño? Well, because it's part of a huge climate phenomenon called Southern Oscillation. Southern Oscillation is an amazing pressure swing formed by a complex self-organizing climate system. The essence of the Southern Oscillation is as follows. Scientists have noticed long ago that a number of various varieties could happen with the weather in the Southern Hemisphere—trade winds can blow or stop, rain can pour in Australia or a hurricane can occur in the Pacific Ocean, snow can calmly fall in the Andes or a snowstorm could happen in Antarctica. And all these things are independent of one another. However, one thing can never happen—that high (or low) pressure is simultaneously set in both Tahiti and the city of Darwin in northern Australia.

What this means is that if the atmospheric pressure is high in Tahiti, it's surely low in Darwin. If it's low in Tahiti, it's surely high in Darwin. And it could be no other way. The distance between Darwin and Tahiti is 5,000 kilometers (3,000 miles)—pretty far apart. But it so happened that Darwin and Tahiti are active centers of the Southern Oscillation, two strange poles connected by a magic weather thread. The magic law is simple: if the pressure drops over Darwin, it grows over Tahiti, and vice versa. That's why this is called Southern Oscillation. The giant atmospheric swing responds to the redistribution of air masses in the entire Southern Hemisphere and acts as its weather kitchen.

Usually, Tahiti has high pressure while Darwin is cloudy. This situation is normal. In this case, trade winds blow properly from east to west into Magellan's sails, anchovy reproduces, and hot sun mercilessly scorches Atacama, the driest and highest alpine desert of the world . . . There is no precipitation in Atacama for many years in a row. A fresh loaf of bread could become a dry crust here in several minutes, and if an old resident was asked when it rained last in Atacama, the old man could easily say that he simply couldn't recall it.

But occasionally, a strange atmospheric incident happens. The giant swing flies up all of a sudden, and then the sky glooms over Tahiti while the sun starts shining over Darwin. And then everything in the Southern Hemisphere goes upside down: in Australia, Oceania, and Indonesia it's dry with blazing wildfires; at the coast of South America, sea critters die; Peruvian army is on the move with heavy equipment; heat-loving hammer fish come to Peruvian coast to great amazement of anglers; and it rains in the Atacama Desert. And the world's driest desert suddenly flourishes with vegetation. But this is a bad omen, because if Atacama is in full bloom, it means everything on the coast is washed away with rains and mudflows and destroyed by hurricanes.

We, too, have a similar pressure swing called North Atlantic Oscillation, which is the weather kitchen for the entire Northern Hemisphere. The magic points of the North Atlantic Oscillation are the Azores and Iceland. TV viewers who like to watch weather reports are used to the phrases "Azores anticyclone" and "Icelandic area of low pressure." Usually, that's the way it happens, clear sky over the Azores, high pressure, and anticyclone. While in Iceland, it's gloomy, cloudy, rainy, provoking a desire to commit suicide. Or at least, to drink lots of scotch or whatever else those taciturn northerners drink at moments of desperation . . .

And then, all of a sudden, the pressure swing displaces the air masses, and everything is put upside down: the air pressure over the Azores and Iceland starts transferring to its antiphase, it drops over the Azores and grows over Iceland. Severe and dry (snowless) winters come to Europe, and Europeans are amazed, being used to mild and high-precipitation winters.

In the last twenty years, pressure over the Azores is abnormally high, Icelanders are suffering from regular headaches, and winters in Europe (including Moscow) are of the kind that we are already starting to get used to—slobbery. Warm western winds bring us greetings from the Atlantic . . .

You could ask a reasonable question: why is it that the western winds bring us warm greetings from the Atlantic and the eastern winds bring us no frosty greetings from Siberia? The reason is that the Earth's rotation causes creation of so-called brave west wind (westerly)—predominantly western winds in the middle latitudes. These constant western winds are in essence similar to trade winds, only they blow in the different direction and at different altitudes. Here's how the system works. At the equator, atmospheric masses are moved from east to west (trade winds). But then the sun-heated air masses rise aloft and move from the equator to areas with lower pressure (45-60° northern and southern latitudes). Due to the Coriolis force, the flows of air masses tend to "bend" westward. That is, the air generally moves north, but its move is vortical rather than straightforward, forming along the way a perpendicular component to its velocity.

The only difference between trade winds and westerly is that trade winds blow into sails, or quite low, while westerly whistles at high altitudes. That's the reason why it takes one hour less to fly from Moscow to Thailand (or from America to Europe) than from Thailand to Moscow (or from Europe to America). This fact amazes tourists greatly. They keep asking one another at various sites about the reasons for this mysterious phenomenon and then start thoughtful and near-physics discussions about the possible role that Earth's rotation plays here, the plane takes off and Earth in its rotation brings Bangkok toward it. And when the plane makes the return flight, Moscow "runs away" from the plane. The others, being wiser, respond—no: firstly, Earth rotates in the opposite direction, and if what you described were true, it would take longer to fly to Vladivostok, but not faster! (*Vladivostok is eight time zones and 6,400 kilometers or 4,000 miles east of Moscow, on the west coast of the Pacific Ocean—Translator's note.*) And secondly, Earth's rotation plays no role here because the atmosphere rotates along with the Earth like one single system, otherwise when jumping we would land in a different place due to Earth's "getting away" from us. And generally, if the atmosphere were not to go along with the Earth in its rotation, it would be impossible to get out into the street, the extremely fierce wind would simply knock you down. And so people are left clueless, "Then what could affect the mystery of planes' speeding eastward and braking westward?"

The paradox of the correct answer is that the culprit here is indeed the planet's rotation! Only it works not directly, but indirectly through the Coriolis force that "bends" the air mass transfer. Generally, if you imagine Earth from the outside, air masses move over its hemispheres as two bagels, or scientifically, toroids. Along the entire equator, the warm air is rising aloft and transferred toward the poles with a slight bend caused by the brave west winds. Closer to the Earth's caps, the cooled air descends. And then it flows back at low altitudes to replace the warm air that's risen aloft at the equator. The low, equatorial part of the geoid is perceived as trade winds, while the north latitudinal as westerly. That's how the atmosphere works in a first approximation. And in a second one, the devil makes its home in the details and particulars of hurricanes, storms, monsoons, winds, precipitation, and other weather features . . .

So, the intensity of westerly depends on the mentioned North Atlantic pressure swing—the more it's inclined toward Iceland, the more intense is the westerly. And vice versa, when the low pressure occurs over the Azores, a stable anticyclone is formed over Iceland, and the so-called unforgettable Great Winter comes to Europe with the freezing up of the Rhine, the Rhone, and the Thames. And these rivers could be covered with ice for up to three months! This kind of severe winter reigns with predominantly northeast winds. And at times like these, it could take as much time for a flight to Vladivostok as for the return flight.

. . . Is anything clear? In this climate thing, there's no making head or tail of it . . .

PART 2

Climate Criminalistics

Pinocchio's age can be easily determined by a saw cut.

Folk wisdom

During military studies at a university, the major explains to the students:

"A machine-gun of an armored personnel carrier can have a maximum vertical angle no more than 30 degrees."

A question from the audience:

"Sir, what are those degrees—Celsius or Fahrenheit?"

The major thinks a little, and then says, "Fahrenheit." A burst of laughter.

"Come on, I was joking," says the major. "It's Celsius, of course!"

A joke

"With this climate, sometimes really mysterious things happen! . . . Do you want your tea real strong or not?"—Klimenko was holding the teapot with brewed tea.

"Medium lousiness, please." I didn't care. "And what was it you said about mysticism? Does climatic mysticism really happen?

"It sure does! Do you know when the three coldest winters were in Russia for the last three hundred years?"

During an exam in heat technology, the examiner asked me whether I had ever flown in an air balloon. This is of the kind of questions that need no answer. Of course, no! I don't know.

"Well, I'll tell you. The three coldest winters in Eastern Europe were in 1708/09, 1812/13, and 1941/42. And it was these three coldest winters that saved Russia three times."

"The God blessed land . . . So, what happened in 1709?"

"The Poltava battle."

"But it was in summer!"

"Yes, Charles was beaten in Ukraine in summer, but he was beaten only for the reason that half of his troops died in winter from cold and hunger. Charles approached Poltava with only a half of the demoralized army. The collection of chronicles describing that winter includes the following: 'That year, Russians were killing Swedes openly and covertly everywhere in homes and on roads, and those who were alive and others were brought to the sovereign, while catching and killing the wanderers, and lots of snow was then and the winter was severely frosty causing many Swedes to die . . .' The end is well known. As to 1812," . . .

"No, don't, I know that sad history very well. No matter how much Stalin wanted to present the old man Kutuzov as a genius commander and Russia's savior, Napoleon's army was ultimately ruined by frost. Napoleon was far from being a fool; he studied meteorological reports of several decades on Russia and knew for sure that severe frost comes in December, while in November temperature never drops below -10°C (+14°F). This suited him quite well. But that year, the frost hit as early as October, and in November thermometer column dropped down to -26°C (-15°F). The result is well-known retreat from Moscow. Napoleon had one hundred thousand troops, but when approaching Smolensk he had only thirty thousand. The rest simply froze to death! And I don't even know what it was related to."

"Maybe, a volcano erupted somewhere?"

"I don't think so. Arctic and Antarctic ice samples show that a very large eruption did indeed happen, but it was in 1808. It's unclear yet what that volcano was, but it's clear that it was somewhere in the tropical area, since the traces of that eruption are found in both hemispheres. Theoretically, it's possible of course, to try to link the 1812 cooling to this eruption. But then what are we going to do with Hitler? With Hitler, something utterly unreal happened! Firstly, it was a calm time with respect to volcanism. Secondly, Germans had at that time the most powerful meteorological service, the best in the world, and it was only natural: they were conducting a huge number of military operations in Arctic and North America where weather forecasts meant a lot. I knew a prominent German climatologist Hermann Flohn, professor at Bonn University. During the war, he was one of the German weather service managers. And I can tell you that Germans were successful in making wonderful forecasts even then!

"So, through 1939, Europe had no cold winters for several decades, and people were getting unused to them. And all of a sudden, two very cold winters struck nearly one after another—1939/40 and 1941/42. Those were the two coldest winters of the century. Neither Hermann Flohn, nor our meteorologists could ever dream of that in their worst nightmare. Both the Russians in Finland and the Germans near Moscow were caught by surprise. Germans were not prepared for such frost; they lacked adequate gear, rifle lubricant, and winter lubricant for tanks ... And now tell me, the three coldest winters ruining three enemy armies—was this not a miraculous coincidence indeed?"

"I don't believe in miracles. So what was the reason for the cold anomaly of 1941?"

"A rare event happened—the North Atlantic swing swung from the Azores to Iceland. Apparently, the same thing happened in both 1709 and 1812."

"Well, let's consider it a coincidence. As the Russian phrase goes, there would be no luck if it were not for bad luck. And let's now go over from mystic coincidences to proletarian regularities . . ."

CHAPTER 1

The Beginning

It's believed that civilization was born five thousand one hundred years ago, with the start of the dynastic period in Ancient Egypt. However, people were living on earth even five thousand years before that momentous event. And quite well, by the way. They were not yet united into states, but were building cities all over. For example, today any tourist entering Jericho is greeted by a poster with a happy message that Jericho is ten thousand years old. And it's true. Ten thousand years ago, the city of Jericho already was. And it was not some godforsaken rural hole, but a real city, or in other words, quite a respectable settlement with buildings of stone and fortified walls.

By the way, in those distant times, there was no Black Sea yet, a small fresh-water lakelet was in its place. And even earlier, in the ice age, a giant fresh-water reservoir spread over the place of this lakelet—the Lake Sarmatia. It was connected with the Caspian Sea and reached the Aral Sea to the east, its northern coast lay somewhere between Saratov and Volgograd, and it reached the present-day Budapest to the west. This giant water reservoir was filled by mighty rivers flowing out of the grand northern glacier.

But when ten to twelve thousand years ago the ice sheet was significantly reduced in size, the Sarmatia Lake no longer was refilled by the mighty glacial streams and almost totally evaporated in a couple of thousand years, leaving behind a small lakelet in place of the present-day Black Sea and two reservoirs separated from one another, the Caspian and the Aral seas. This lake was separated from the Sea of Marmara by the dry Bosporus Isthmus—the World Ocean level was then about 40 meters (130 feet) lower than it is today. So what then caused the Bosporus "dam" to become the Bosporus Strait? Why did the ocean water level jump so high, pouring into the Black Sea basin?

The point is that the ice sheet melted unevenly, first, the European glaciers were gone, then the North American ones. The last one to melt was the so-called Laurentian continental shield, the shelf ice with its center located at the present-day Hudson Bay. When eight thousand years ago the cofferdam, separating the glacier from the ocean water, collapsed, the Laurentian shield began its rapid destruction and was destroyed in merely two hundred years. Catastrophic speed! And it had twice the amount of ice of present-day Greenland. That amount of ice was enough to cause the World Ocean water level to rise by 7-9 meters (23-30 feet). And then the water came running down through the Bosporus channel into the future Black Sea, creating a waterfall that was ten times as powerful as Niagara. Oh, how magnificent it was! The water rose at a rate of almost 20 cm (8") a day or 6 m (20 feet) a month, so the Black Sea was filled up to present-day level in just two years. People who lived on the coasts had to move out, away from the water that was approaching in front of their eyes. Thus, about eight thousand years ago, big diasporas of people were formed, who carried with them into the future the legend of the Flood (details of those tragic events are described in my book *Civilizer's Fate*).

Before we get into further discussions on human history through the prism of climatology, we must make a small but wonderful digression. However, I would not call it a lyrical one . . .

Let's speak about dating. When climatologists, archeologists, or historians state that an object they found is, let's say, seven thousand years old, what do they mean? What do they use to determine the age of that object? Historians often relate to time with help from physicists, and then they relate to so-called radiocarbon age, or the age determined by radiocarbon analysis. But the problem here is that practically not a single historian knows that radiocarbon age differs from the calendar age, and this difference could be quite significant. For instance, if a sample is five thousand years old on radiocarbon scale, it is six thousand calendar years old. And the difference of thousand years in the era of Ancient Egypt is quite significant, when a historian in this case could risk confusing Ancient Kingdom with the New Kingdom.

There's a scary suspicion that 99 percent of historians, fortunate to read this book, would be shocked by this discovery. So, for their sake only, I shall discuss the radiocarbon method in some detail. This is useful even more so as climatologists also widely use this method and many of them, too, don't know about the difference between the radiocarbon and calendar ages.

Willard Libby, who in 1940s invented a method for radiocarbon dating, received the Nobel Prize that was well deserved—it turned out to be a very convenient thing. And most importantly, even a homemaker can understand how this method works (if, of course, she studied in a Soviet but not an American public school). And thus, it would be a sin not to remind the broom—and pot-laborers about the essence of the method.

So, you can write it down. Carbon in the atmosphere is present mostly in the form of its dioxide—carbonic acid gas. But besides the normal carbon ^{12}C, atmosphere also contains some fraction of radioactive isotopes of carbon—^{13}C and ^{14}C. Half-decay period of ^{14}C is 5,730 years. A reasonable perplexity: with such a short lifetime, all carbon isotopes should have completely decayed long ago, but since they are still present, it means they are coming from somewhere. Correct: under exposure to cosmic radiation, carbon isotopes are constantly produced in the upper atmosphere from nitrogen atoms. Thus, there are no problems whatsoever with replenishing the atmosphere with carbon isotopes, so homemakers should have no reason to worry.

Any living organism breathes and feeds, or in other words, exchanges carbon with the environment. And when the organism dies, it stops replenishing its reserves of carbon and thus fixes in itself the content of ^{14}C. After that, all accumulated amount of this isotope can only decay. We know the half-decay period. We also know the content of the isotope in the atmosphere. Upon determining the amount of ^{14}C in the found piece of wood, it's possible to learn when that tree was sawed. If the ^{14}C content in the sample is exactly one-half of its content in the atmosphere, it would mean that half-decay period, or 5,730 years, have passed since its death. A very simple exponential dependence. Tip-top!

The only problem is that the content of ^{14}C in the atmosphere at different periods in history isn't the same! It depends on climate changes, on ratio of areas of dry land versus ocean, on solar activity, on parameters of global carbon circulation (how actively carbon is extracted from the atmosphere by ocean and land flora and fauna) . . . All these fluctuations cause the relationship between residual ^{14}C and time to be not as simple as on the half-decay graph. That's why the radiocarbon time can differ greatly from the calendar. For this reason, special correction tables were created to convert carbon time into calendar time. These tables are created by a "mind-boggling" number of laboratories—about a half dozen around the world. It's quite a complicated and tiresome process. Once every five to six years, correction tables and calibration curves are reviewed and updated. But for some reason, not many even in the scientific world know about them.

Calibration curves have a very quaint form. In order not to scare away readers visiting a bookstore and leafing through this book, I'll try to include as few graphs and formulas as possible, since every formula is known to reduce the number of readers by half. So I'll just explain in words that, for instance, a real calendar date of 299 BC has three matching radiocarbon ages—2171, 2200, and 2254 carbon years ago (I should note that the starting point for radiocarbon age is traditionally considered 1950). Now let's assume that a radiocarbon age of 2450 years matches a range of calendar years, 343 years wide (from 757 to 414 BC). If we consider the error of the measurement instrument 2-3 percent, the calendar range width increases on both sides by another fifty to seventy years! In other words, if radiocarbon analysis shows a historian that a sample is 2,450 years old, then the year when Rome was founded could be easily confused with the end of the Peloponnesus War.

Moreover, the "distance range" of the radiocarbon method is limited by a relatively short life-span period of the carbon isotope. In 5,730 years, half of the ^{14}C decays, in the next 5,730 years—half of the remaining half, or three quarters of the initial amount. And so on. The more time has passed, the less number of atoms are available for researchers to work with. And the sample had scanty amount of the isotope even at the start! And the sensitivity of physical instruments is not limitless!

The biggest enthusiasts of the method say that it can produce good results up to 40,000 years back. The others believe that reliable results of radiocarbon analysis are within the range of 10,000 years and not deeper. We shall not argue, but just say that the correction tables that are being constantly updated are divided into two parts. The first part ends with a date of 7,210 radiocarbon years corresponding to 9,000 calendar years back. At this range, the error is smaller, but still can reach 200-300 years. The second part of these tables is up to 22,000 years. And here the error is plus or minus 1,000 years, which is totally unacceptable for historians. But they have no reason to look into such depths—history has started literally only yesterday. So, the "distance range" limitations of the radiocarbon method concern not historians but rather other scientists. Like paleoclimatologists, for instance . . .

And there's another snag in the radiocarbon method—the human factor. There are not too many labs in the world capable of professional radiocarbon dating, but far too many people eager to make money from it. Dating of a sample costs about a thousand dollars. Not sour dough, you would agree. Russia has about twenty labs that would gladly take your money, but you could trust the results of only three or four of them.

The lab equipment for radiocarbon analysis is not just very expensive (about one million bucks), but very complicated, too—it's no simpler than a jet plane. And it looks quite impressive: a forest of tubes, several tons of metal, vacuum pumps, mass spectrometers, cylinders with super-pure helium, argon, nitrogen . . .

As of today, the laboratory of radiocarbon dating in the Geological Institute of Russian Academy of Sciences is considered one of the best in Russia. It is headed by an amazing man with an amazing name and amazing fate—Leopold Sulerzhitsky. The world has many specialists in natural and technical sciences who came to work in areas of humanities (like, for instance, this author), but the reverse practically never happens—for a humanist to start working in an area of technology. And this is natural: it's always possible to jump from complex to simple. If you have a good educational foundation, you can later do whatever you wish—any jabberwocky, but try to employ a humanist at a factory as an ordinary engineer . . .

Leopold Sulerzhitsky is the rarest exception. Having graduated from a conservatory as a cellist, he gave up this worthless activity and became a real man—a scientist. Today he is over sixty and leads a first-class laboratory. He is the only head of a laboratory in Russia without higher education (conservatory does not count, of course). To grant him a title of a Senior Scientific Fellow, a special decree was required by the presidium of the USSR Academy of Sciences. And such decree was issued, which does credit to the Academy. Sulerzhitsky is a true fanatic of science. He needs nothing except science. He has only one suit that he wears when going on expeditions, or to work, or to a farmers' market for potatoes. Isn't he a saint?

About ten years ago, Sulerzhitsky was working with relish on the Wrangel Island. Remains were found there of mammoths that, as it turned out, lived just about 2,500 years ago, in times of Rome. About a couple of years later, *Nature* published an article about this sensation. *Nature's* publication had eight signatures except that of Sulerzhitsky, although it was Sulerzhitsky on the Wrangel Island, who picked at those mammoths and did the dating. When his colleagues became indignant with this fact, Sulerzhitsky just gave it up as hopeless. He was not looking for earthly fame—he was interested in the process. Clearly, a saint . . . By the way, if a novice climatologist brought a sample to the noble Mr. Sulerzhitsky, it would not mean that an adequate result could be produced. The reason is that the lion's share of success depends on how properly the sample was collected. It's easier in this sense for an archeologist, who finds a ship fragment or a wooden shield, drags it to a decent lab, and gets a decent result. But for a climatologist, an error in selecting a tree could be fatal. Not all trees are created equal in usefulness. And sample collection here lies far beyond the limits of professional skills—almost in the area of arts.

. . . Now, wait a minute. Why a climatologist would even need radiocarbon analysis of any trees?

You're right—this also needs an explanation. For if I give you no explanation but just make a statement that, for example, the intellectual spike of the Axis Age—the turning point in the entire human civilization—was caused by a powerful cooling, then some former but proud student of the History School of the Moscow State University would definitely ask, with his lower lip scornfully protruded:

"And whatever gave you the idea that it was global cooling then? How could it be known in general, when and how much the world average temperature rose or dropped? Back then, they had no meteorological stations, computers, or world-wide weather service."

Right on the money—they had none! It wasn't for nothing that this person studied history in the university. So, the explanation is needed . . .

There's such a thing—criminalistics. The science about traces. It's about how, using minute scratches and with no witnesses, to restore the picture of what happened. Paleoclimatology is the same criminalistics, but it tracks just one "criminal"—the climate. And it has a number of methods to catch the escaping fugitive. Let's glance at them—solely for the purpose of general education. And those who are too lazy to dig through the rubbish of past eras could go directly to adventures, or to the Part Three of this great book.

CHAPTER 2

When Trees Were Small

For obvious reasons, it would make sense to discuss only those methods of climate reconstruction that would enable to restore the climate picture for the last several thousand years. Therefore, we're not going to discuss geological and microfaunistic methods: the former deal with the range of tens and hundreds of millions of years, while the latter with tens and hundreds of thousands of years. Microfaunistics studies microremnants of ancient fauna in sedimentary rocks, as a rule in bottom sediments of lakes, seas, and oceans. The sediment accumulation rate is extremely slow, thus one small sample could contain information about thousands and tens of thousands of years. It's very inaccurate, and we would like to have something more precise—methodology for temperature reconstruction should produce not a century-average value but ideally an annual or even better a seasonal resolution, so we would be able to say, "Aha! The winter of 1319 was colder and its temperature was 1.5°C (2.7°F) lower than our today's norm."

And science does have such methods.

Dendrochronology

In the early twentieth century, German-Russian climatologist, Vladimir Petrovich Keppen (born in St. Petersburg, but later moved to Germany) said that vegetation is crystallized climate. The statement was very apt, everyone liked it, but it should be said in all fairness that ancient Greeks surmised as much. They realized that whatever is the flora, the same is the climate. Or rather vice versa, whatever is the climate, the same is the flora.

And whatever is the flora, the same is the fauna.

And whatever are flora and fauna, the same are humans, since climate, flora, and fauna utterly determine living conditions for humans in a given area. Actually, this fact should be clear to any normal person, that it would suffice to place side by side a Greek, a Papuan, and a Chukchi in order to see first-hand that living conditions utterly determine the human way of life (clothing, traditions, moral) and appearance. Trees are the living chroniclers of an era. Change of seasons is reflected in rings on saw cuts. A new ring is formed every year. However, not all trees are so diligent—tropical and subtropical trees have no pronounced rings because there is no sharp change of seasons in those areas and hence winter does not leave a distressing scar on the tree's long-suffering body.

California sequoias are best suited for studying tree rings. Firstly, they have very thick trunks, making it easy to study the rings—each ring is a finger-thick. Secondly, sequoias live thousands of years. It's a shame, of course, to saw down such a tree—thick as a house and a contemporary of George Washington—but studying its saw cut enables to look hundred years back. By the way, today, it's not necessary to saw trees down, for, sophisticated western science invented a nontraumatic method of obtaining needed information by using special thin drills that can extract a sample without killing the tree.

So what do these rings show? The rings can show how the tree was doing in a given period—fine or poor. If the tree was fine, it thickened fast and its rings came out wide. But if the tree lacked something, the annual ring would be narrow. But what did the tree lack for healthy development—was it heat or moisture? That is, was that period cold or dry? This is the question, which dendrochronology method cannot answer. And that's only the first demerit of dendrochronology.

The second demerit is that the ring width gives an idea of climate conditions for the vegetation period only, that is when the tree grows. In Arctic, for instance, the vegetation period lasts only two months, while in the south it's longer. But in any case, although the ring is called "annual," it contains information only about what kind of summer was in that year. Specialists generally know this but eagerly forget it and try to interpret the data on obtained temperatures on the basis of studying the rings as annual averages. This was how a scandalously sensational publication of American scientists was born in 1998-99.

It was a bomb! Americans decided to reconstruct the average global temperature of the last one thousand years using only dendrochronological data obtained, by the way, only at high latitudes. And they determined that since the start of the twentieth century, the planet experiences a totally unprecedented warming exceeding in scale all what happened during the period they were reconstructing. It's absurd, and we'll prove it later. It's a classical example of how bad an error could be made when ignoring the method's limits.

There's another issue to be noted here. Mathematical processing of the dendrochronological information obtained is so complex and multistep that researchers are left with only climate events of several decades on the timescale. Centennial fluctuations are cut off during mathematical processing. The most experienced and honest dendrochronologists write about it directly, albeit too humbly: three lines in a twenty-page article. Like "harmonics are cut off during filtration," which are the harmonics that are cut off, very few nonspecialist readers can understand that. Nevertheless, we must learn the method's another demerit—even in dendrochronology's longest series (several thousand years), it's impossible to see centennial or the more so millennial fluctuations: "harmonics are cut off."

On the other hand, dendrochronology has its advantage: the method provides annual resolution. That is its indisputable merit. Here's an example of one of correct dendrochronological studies.

In Russia, sequoias unfortunately don't grow. Hence, the scientists from the Institute of Plant and Animal Ecology of the Urals Division of Russian Academy of Sciences (Yekaterinburg) have thought long ago to use Siberian larch for dendrochronological studies. This wonderful tree enabled to look four thousand years back and mark the years of extreme climate events.

The choice of larch might seem strange: what kind of rings would be on a tree that grows on the edges of Western Siberian tundra? Indeed, very thin. But larch has an advantage, too. As most trees grow older, not only their trunk diameter thickens, but also their bark, or thermal insulation. Therefore, a thick tree may not feel the strike of elements, saw-cuts of big trees have fewer rings damaged by summer frosts (so-called frostbitten rings), while larch with its thin deciduous bark meticulously signals to scientists about each weather mess. And though its trunk diameter is small, so is its bark thickness that rarely exceeds 3-5 mm (0.1-0.2").

The study of frostbitten rings enabled to select the years when at the Polar Urals the temperature in summer fell below 5°C (+23°F): 1466, 1573, 1601, 1708, 1783, 1797, 1811, 1857, 1862, 1872, 1882, 1891, and 1968. And the severest frosts happened in 1601, 1783, 1857, 1882, and 1968. (The year 1601 is already familiar to us. The frostbitten rings corresponding to this year are also present in pines growing in North America. The year 1783 is also familiar to us. The dry fog lasting from May 24 through October 8 that year covered the area from Norway to Syria and from England to Siberia. In Russian capital in midsummer, according to contemporaries, "the sunlight was weaker than the light from the full moon." It was the result of the activity of the Icelandic volcano Laki. We'll probably talk about the rest of the years at an opportune time.)

It's not enough to choose a good laboratory for sample analysis—the samples themselves have to be selected properly. In the Urals, they selected their samples correctly by choosing stand-alone larches. This is essential. It's not important what you're dealing with—bushes or trees, but it's necessary for sample selection to choose only trees that stand alone at the border of their community—at the edge of forest-tundra, forest, or alpine meadow: they sense the strikes of fate sharper. In general, the marginal trees are needed, the ones that depend less on their community and more on the climate. Their signal is brighter.

About twenty strict rules for sample selection are honestly described in special brochures published on newsprint in about two hundred copies. Nobody reads them except fanatics. And it's a shame, for as we have already mentioned, an incorrectly selected sample plus a crappy lab could kill any work. And since there are much less fanatics and good labs than there are curious researchers, 70-80 percent of dendrochronological work could be used as toilet paper only.

Palynology

This method got its name from the English word *pollen*. Palynology enables to reconstruct temperature and precipitation using fossil remnants of plant pollen and spores. Why is this possible?

Science since long ago has a two-parameter Holdredge diagram. Its vertical axis shows annual average temperatures, while the horizontal axis shows annual average precipitation. The diagram itself has delineating lines of all the vegetation communities known to humans—Arctic tundra, taiga, broad-leaved forest, desert, semidesert, forest-steppe, steppe, savanna, tropical forest . . . That is, knowing just two parameters—the annual averages for temperature and moisture—enables to determine the particular zone one is dealing with. If, for example, annual average precipitation is 600 mm (24") and annual average temperature +5°C (41°F), that's mixed forest in temperate zone, like Moscow. Precipitation 4,000 mm (160") and temperature 27°C (80°F) point to tropical rain forest. Temperature 5°C (+23°F) and precipitation only 200 mm (8") point to Arctic desert.

In other words, if somewhere in nature we find preserved residue of plant pollen capable to time-date, then we have a true paleothermometer! Peat bogs are excellent storages of such residue. As you undoubtedly remember from school lessons in natural science, peat is unfinished coal. It consists of 98 percent plant residues and 2 percent remnants of animals and microorganisms. In essence, peat is dirt. This dirt is of special value due to the fact that peat accumulation rate is very high—it can reach several millimeters (0.1-0.3") a year. No comparison to geological sediments! Extracting peat cores from a bog gives us detailed information—literally by years.

But just as in dendrochronology where not all trees are suited for paleoclimatic samples, the same goes for palynology where not all bogs are suited for core extraction. Only high moors! Eutrophic swamps are of no interest to us because they accumulate water from the bogs located higher, which messes up the whole picture. Researchers need only those bogs where the water drains from, but not into.

Next. Bogs located close to humans are utterly wrong for research. Instead, only those bogs are good that are located as far as possible from any objects of economic activities. The point is not that "civilization contaminants" can get into bogs, like petrochemical or other chemical products—that's not too scary. What's worse is the biological contamination. If a bog is located within one mile from fields or orchards, it means that in the last several hundreds or thousands of years, all the peat strata would be full of hindrance—the pollen of cultivated plants.

For these reasons, palynologists get into thicket of tropical forests, Siberian backwoods, and tablelands of Patagonia. By the way, Russia is an excellent place for palynologists, because it's almost entirely a continuous godforsaken place—from the Chukchi Peninsula to Belarus and from the Taymyr Peninsula to the Northern Caucasus.

After palynologists extract peat cores from a bog, they study them by strata: they look to what the plant pollen is contained and in which strata. And they don't look for pollen of something specific, like oak or dandelions, but select multi-species variety because only dozens of species provide the full picture. For instance, if herbaceous communities prevail over arborescent-shrub species in peat strata, it means some time ago steppe was here. So I hope the principle is clear.

Research shows that at any particular place, dramatic changes happened in tens of thousands of years. For example, in Central Russia, landscape changed from tundra or even Arctic desert to broad-leaved forest. And then the ratio of pollen of different species is used to determine annual average temperature and precipitation. And the accuracy of this method is half degree (1°F) for the annual average temperature. Not a bad result!

When Klimenko was reconstructing the climate of central part of Russia, he himself appreciated palynology by analyzing peat cores of the Western Dvina and Valdai bogs. In his works, the detailed climate history of annual and seasonal average temperatures was reconstructed, as well as the annual average precipitation for the last five thousand years. So what interesting facts were revealed?

It turned out that the winters of 1990s had no analogues in the five-thousand-year climate history. They were very warm! And the summers were ordinary—much warmer summers than in 1990s happened many times. As to precipitation, it was within the limits of climate norm—no sensations here. This proved once again that the global warming affects mostly the high latitudes and not the summer but the winter temperatures. And it's quite beneficial for us, my dear readers. But we shall talk about it in due time. And in the mean time, I would like—just for the like of it—to cry over Russian destitution . . .

Due to the fact that Russian science has not enough money, it lacks storage facilities for samples and the latter are discarded upon completion of studies. In Russia, the evidence of past millennia is not protected. In the rest of the world, special storages exist for scientific samples, while we in Russia have a generous soul: we pour out the peat dirt into a ditch—and the sample is no more. And that's a shame. In the West, new generations of researchers occasionally get back to old samples and—occasionally—make new remarkable discoveries when using them, because new research methods appear all the time.

I recall when Klimenko first told me of the impossibility to store samples in Russia, I heatedly exclaimed:

"But then we should give them to foreigners as presents!"

"That's what we're doing. In the last ten to fifteen years, scientists from various countries began pouring into Russian Arctic in crowds. But in the Soviet times, entire Arctic was closed. In August 1990, I came to the Dixon Island on board *Anton Chekhov* ship. As you know, the Dixon is extremely far away from all the borders. Thousands of kilometers (miles) in any direction: 3,000 kilometers (1,900 miles) to Norway, 6,000 kilometers (3,700 miles) to Alaska. It would be generally hard to find in the world a place more remote from any borders. And what do you think, who was the first to board that ship? Border guards! Simply a madhouse! The entire Soviet Arctic was a continuous border with who knows what. With polar bears, perhaps. So when this madhouse finally ceased to exist, scientists from all over the world happily started coming to Russia to conduct research. They grabbed and took out samples of everything they could take and already filled all their storages with them."

"Oh, how nice!" . . .

Glaciology

Glaciology studies permanent ice. And it's not necessarily in Arctic, Antarctica, or Greenland. Permanent ice is plentiful, right in our backyard—in Europe. It's in France, Switzerland, Iceland, Germany . . . While writing this book, this author almost went to one of Austrian glaciers—not to drill cores, of course, but to ski a bit. But then decided that in Bulgaria, it would cost half as much with the same result, and so went to Bulgaria . . .

And now, upon returning from Bulgaria, I continue working for the eternity by telling you about permanent ice and European glaciers. Mountain glaciers react sharply to fluctuations in annual average temperature. If it becomes slightly warmer—the glacier's lower extremity, thaws and goes higher in the mountains. When it turns colder—the glacier drops its tongue lower.

The glacier edges could be dated easily by the felled trees. The growing glacier works like a bulldozer: it cuts trees. And trees are a convenient research object: the age at which the tree died is easily dated by its rings or by radiocarbon method. How to determine the time when the tree died by its rings? A volcanic (or a frostbitten) ring has to be found—it's usually very thin. For instance, on saw cuts of American sequoias or European oaks, it's easy to find rings corresponding to Tambora eruption of 1815. The ring count from 1815 toward the bark would easily tell the year when the tree was cut by the advancing glacier. After that, it's easy to determine the temperature for that year. How? Very simple! It's known that temperature drops 6.5°C (11.7°F) with each kilometer (3,300 feet) of altitude. In other words, a temperature drop of 3°C (5.4°F) is equivalent to an ascent of about half kilometer (1,650 feet). Or if temperature in the Alps changed by one degree (1.8°F), the glacier moved 160 meters (525 feet), that is vertically. As to its horizontal movement—it could be hundreds and thousands meters (up to a few miles). You have to be blind not to notice if a glacier advanced by a kilometer (0.6 mile)!

The Alps have some very beautiful glaciers that people paint century after century. And it's even possible to make climate conclusions based on those pictures and engravings. They all have written on them the year of their creation, and it's always possible to find in them some benchmarks (a church, a cliff) serving as reference points regarding how much the glacier has moved by now in comparison with the time when the pictures were created. Of course, pictures and engravings are not micrometers, but we don't have to catch every millimeter here: an error of 50 meters (160 feet) corresponds to an error of just one tenth of a degree (0.2°F).

But we digressed slightly from glaciology to the area of historical climatology—the method enabling to restore the climate of the past by using evidence of chroniclers and descriptions of contemporaries, but it's not terrible since both methods often go hand in hand. Let's look at the famous Swiss glacier Aletsch, the most researched and scrupulously dated glacier in the world. Around it, such an incredible amount of cut trees were found and such a huge number of historical chronicles were read that when one and the other were matched it became possible to restore temperatures in this region very accurately for almost three thousand years.

However, the version of glaciology more common to the public is the isotope one. Everyone heard that scientists drill Antarctic ice, extract cores, and exhaustively study them in their laboratories. Even the then US vice president, Albert Gore, flew to Antarctica to see those studies—and got very impressed. It's very easy to impress an incompetent person interested in where the taxpayers' money goes . . .

By the way, some stuff they have is indeed worth admiration. An ice-core sample is only about 20 centimeters (8") in diameter, while its length measures up to 4 kilometers (2.5 miles). It is cut during extraction and each piece is handled carefully so it's not soiled with bore lubricant or cracked, because if air gets inside the ice core, it's as good as lost. An incredibly complex operation! Processing the samples is no less complex. And by the way, they are processed not in Antarctica. The ice cores are brought for processing to Grenoble (France) or Columbus (Ohio, USA). They are transported over huge distances in special refrigerated chambers and upon completion of studies placed in refrigerators for eternal storage. The planet's ice sheets are drilled for decades, and all cores extracted during all those decades are saved. It's indeed costly. But science requires sacrifices from taxpayers; otherwise, what's the good of having taxpayers in the first place?

In the labs, scientists use mass spectrography to determine chemical and isotope composition of cores. Through the ratio of oxygen isotopes ^{18}O and ^{16}O in atmospheric precipitation (or in snows that fell in Antarctica and compacted into ice), scientists determine the past climate. The point is that this isotope ratio correlates clearly with temperature.

Obviously, like in any business, ice sample selection also has its subtleties. Ice is a so-called rheological liquid. That is, a liquid that is not subject to Newton viscosity friction law. It flows under different laws. Layers of ice can fold into creases, crumple, and get under one another. And so it's possible to find younger layers under older ones in the same core. And not knowing this could risk obtaining wrong results. It should be noted that correct determination of a drilling spot is a study in itself. Similar to bogs, the drilling is done on ice domes, from which the ice could only flow down.

Above-mentioned Albert Gore brought from the ice continent one simple idea (American liberal democrats in general never have complex ideas) that the content of carbon dioxide in the atmosphere correlates well with the warming of the atmosphere. He was told this in Antarctica. This idea came into Gore's head so firmly that since then he started fighting strongly against global warming, for Kyoto Protocol, and for smothering worldwide industry that actively emits greenhouse gases.

Occasionally, tiniest air bubbles can be found frozen-in into ice cores, and studying their gas composition enables to reveal the composition of the Earth's atmosphere in deep past. Actually, carbon dioxide content in the atmosphere correlates well with the temperature—high peaks of temperature on graphs coincide with high peaks of CO_2 content in the atmosphere. That is, there was more carbonic acid in the atmosphere in the eras of warming than in the eras of cooling. But where did it come from, if millions of years ago, there were no actively smoking factories or plants? Somehow, Gore never thought about it. But it came from the oceans. First, the average global temperature rose, and then the greenhouse gas concentration in the atmosphere increased: due to the rise of temperature, oceans began to actively emit gases, like warm champagne. But Albert Gore never learned this. His hard drive had no space on it to record this information. And so he continues up to now to be a fearless fighter against carbon dioxide . . .

Limnology

Limnology is analysis of lacustrine (or lake) deposits. Lakes are true accumulators of climate information. There are three kinds of analysis of lacustrine deposits—diatom, microfaunistic, and isotopic.

Very wise people like me, when hearing the word *diatom*, realize it's something consisting of two atoms. And that's where they err! Diatoms are a class of algae. There are about five thousand species of diatoms. And the ratio of these species in water reservoir depends on the temperature. Some species develop well in cold water, while others in warmer water. If water heats up by a degree (2°F), the species' composition changes. It's a living thermometer! By studying diatom mix, it's possible to determine what the temperature in this or that era was.

Another determination could be made by just the nature of lacustrine deposits. For example, if much sand is found in some layers of a dirt core from the bottom of the African Lake, Chad, it means the Sahara was advancing at that time. If there's no sand at all but there's tropical microfauna, that's a sign of tropical forest profusion. So, the sample-core layers one after another tell us about climate changes in the given part of Africa. It's cool.

The deficiencies of this method are the same. Firstly, the proper lakes should be chosen—oligotrophic (high) ones, or the ones where the water flows out rather than flows in and confuses the picture. Secondly, there are complications with dating. The layers are usually dated by the use of the standard radiocarbon method with all its inherent deficiencies. For an age of two to three thousand years, the error is plus or minus one hundred years. For the age of early Christianity, it's plus or minus fifty years. But for the last five hundred years, the use of radiocarbon analysis would be generally senseless—the spread is too large leading to an error by three hundred years. There are two basic reasons for this: high slope of the calibration curve (it's an inherent feature of the method) and accumulating distortions in the natural course of ^{14}C concentration in the atmosphere due to human activities (for example, concentration level of ^{14}C in the atmosphere increased twofold since mid-1950s as a result of nuclear arms tests).

So all those pretentious "scientists" and spiteful creationists who, while gibing at the radiocarbon method, cite funny examples in their books and brochures—like when a live mollusk was taken and then dated by the radiocarbon method determining the mollusk to be dead one thousand years ago—they just don't know the limits of the method applicability. They prove with this not the loss of a thousand years of history or existence of God, but their own narrow-mindedness . . .

A similarly dark period for radiocarbon analysis is the latter half of the first millennium BC—antiquity. In that period, if the radiocarbon analysis is used incompetently, it's easy to confuse the era of the fall of Babylon with the time of Hannibal's campaign.

Therefore, when doing climate reconstruction, it's best to use not one, but several dating methods. For instance, when dating layers of lake deposits, it's possible to relate to external indicators. It's known for example that a volcano that erupted in 1552 and 1763 is not far from the Lake Chad. So one has to find two layers with volcanic deposits that provide two reference points, and then, with the knowledge of deposit accumulation rate, just count the layers . . . Similarly, the volcanic layer of Tambora (1815) can be well seen and for this reason, used for dating in lake and sea bottom sediments on the border between Indian and Pacific oceans.

An indicator could be not just a volcano, but also any other historic or geophysical event. Or a structure even! For example, no lake in the world has a nonfluctuating water level. At times, the range of the fluctuations is simply striking. Let's take the well-known Dead Sea. Today its surface is 400 meters (1,300 feet) below sea level. But on the walls of Qumran monastery, located at quite a distance from the Dead Sea, the preserved lake-water marks made by the monks show that in ancient times the water level rose by 40 meters (130 feet).

Another example. The largest lake in the world, the Caspian, is also subject to mind-boggling rising and lowering of its water level. The witnesses to this are remnants of big cities and ports that lie today submerged on the Caspian bottom. For instance, the island-city Abeskun at the southeastern part of the sea has been totally submerged. And scuba divers in the neighborhoods of Baku can observe ancient city blocks and architectural monuments on the sea bottom . . .

Current level of the Caspian is below its mean value. But there were eras when many kilometers (miles) of its bottom were exposed. The wonderful and majestic indicator of Caspian fluctuations is the Derbent fortress wall. Derbent is generally a very remarkable city, being much older than Rome. The place where it's located was called since antiquity the Gate of Peoples. It is here, in the narrow pass between the Caspian coast and the Caucasus ridge, a convenient way lies from Central Asia to Europe. As you understand, if a place is popular for passage, sooner or later someone would establish a store or a fence in this place—to collect fees for passage. That's how the city of Derbent was born, which in the past was the capital of a powerful state. The Great Derbent Wall, 18 to 25 meters (59 to 82 feet) high, 5 meters (16 feet) thick, and 75 kilometers (47 miles) long, extended far into the depths of the mountain ridge, blocking the free passage from Europe into Asia and back.

Roman Empire and Persia fought desperately for centuries to capture Derbent. Any conquerors—Huns, Avars, Mongols—who went west from depths of Central Asia, passed through this place. They went "upward" and then turned "left" toward Europe. It was impossible to bypass the blessed Derbent: on one side, Caucasus was standing like a solid wall, and on the other side was the sea. To skirt the Caspian from the east? It was not convenient: the full-flowing lower Volga River had to be crossed, as well as snow-covered steppes or waterless salt marshes. And then, what is there to do in these boring desert places? At the same time, people lived in the Caucasus foothills from time immemorial and accordingly there always was someone whom to trade with or rob. It was good to pass through Derbent!

Those were the reasons why Derbent had cyclopean fortress walls—long, high, double. They protected the city from the enemies moving from both north and south. And naturally, they blocked the movement of those caravaneers who aimed at slipping through without paying duties. A busy seaport was located between the walls, whose entrance was also reliably secured with huge chains.

All famous travelers went through Derbent—Ibn Fadlan, Al-Istahri, Afanasy Nikitin, Jenkinson . . . And on their way they all left description of the mighty Derbent walls creating a fantastic impression upon all travelers. Remnants of those walls remain even today, their ruins still creating indelible impression on all visitors, but for a different reason—the walls go down 300 meters (1,000 feet) under water. This puzzles visitors: "Did they build walls under water in ancient times?" Of course, not—that was the sea edge when the walls were built.

Thus, information on lake water level is the key element in global paleoclimatic picture. Man-made walls are not everywhere to enable tracking the water level fluctuations, but if trying hard it's always possible to find some traces of long-gone water. For instance, lake terraces of African Rift Valley with traces of water levels enable doing this with ease. Only it would be wrong to think that the higher was the lake-water level the greater amount of precipitation fell caused by warm climate. It's not necessarily so. Perhaps, it could have been the other way around—much colder and thus less water evaporation from the lake. Therefore, to get a correct answer to the question of climate, various paleoreconstruction methods should be combined.

Uranium-Thorium Method

Compared with the radiocarbon method, this method is simply remarkable: it has smooth calibration curve with no bends, enabling to obtain dating that's close to the calendar. But unfortunately, one analysis costs ten times more than also-not-so-cheap radiocarbon method. And it's used in geological scale of times.

Historical Methods

If back in the past we had criminalistics, today we have only "witnesses' testimonies" . . .

Historical climatology is reconstruction of a picture based on documents, such as chronicles, testimonies of ancient historians, works of art, and since the sixteenth century—logbooks and travelers' diaries . . .

Dutchmen observed the weather thoroughly since the eighth century, since Holland is a country of windmills and water mills whose performance depends directly on the weather—whether or not would the wind blow or the water freeze. And the transportation issue played a role, too—Holland is a country of canals, and commerce depends on transportation ways. So, generally speaking, they were observing the weather.

In Europe in general, many liked to do weather observations—monks, pharmacists, doctors . . . There are simply shocking examples of selfless devotion. For example, one doctor in Iceland in the early eighteenth century in the course of forty years—up to his death—conducted meteorological observations several times each day, for which he bought some expensive (for those times) meteorological equipment in England. So he should be the one to be canonized! Along with Sulerzhitsky.

Using old chronicles, paleoclimatologists composed quite good seasonal reconstructions for Germany, France, and Holland. A little later, reconstructions appeared for Hungary, Bulgaria, Portugal, and Spain. Retroforecasts came out quite good—for all seasons, with accuracy up to several tenths of a degree. It's clear that reconstruction accuracy depends on density of observations and number of sources. In places where civilized people lived since long ago, reconstructions go deeper into the past. For instance, in certain parts of Holland, climate reconstruction was possible to start back in 764. For some areas of France, Poland, Czechia, and Germany, it's done starting in 1000, while for most other regions of Europe it starts in 1500.

In Russia, similar works are conducted, too. But with great difficulties! Russian scientists attempted to reconstruct climate fluctuations in the basins of the Kara and the Barents seas for the last five hundred years. There are diary records of the very first English and Dutch expeditions of mid to late sixteenth century. There's vast amount of material for the nineteenth century—in some of its years, up to hundred ships annually visited the Kara Sea. The logbooks are available—it would be sinful not to use such a database.

Many have read and many more have seen movies (there were two) or musical based on Kaverin's novel *Two Captains*. It was a very popular creation! Poetics of the north. Explorers of Arctic . . . Enthusiasm of Arctic exploration under Stalin was such that it held in Russian culture for several decades even after the death of the "great leader." But think about it—why all of a sudden did the Soviet authorities decide to explore the north? Well, it's clear—all sorts of mineral deposits, the Northeast Passage . . . But besides the desire, there also should be the possibility! The possibility for keen Arctic exploration came through an unexpected warming.

The older generation remembers this term, *Arctic warming*. Back in the USSR, it was used as often as *global warming* is used today. For some reason, Arctic climate in 1930s-40s indeed sharply improved. For example, winters on Spitsbergen became 8-9 degrees (14-16°F) warmer.

However, the holiday did not last long: just as it warmed up in 1930s-40s, it cooled down sharply in 1950s-60s. But that warming enabled to develop the Northeast Passage and to build numerous cities, towns, and settlements on the dismal boundless expanses of Soviet Arctic. And among those cities were such polar giants as Norilsk.

In spite of the fact that the reasons for the Stalinist warming of Arctic were not fully understood, it's known today that it was not unique. Something similar was repeated three to four times in the last five hundred years. It's clear from all of this that the Stalinist Arctic "thaw" had nothing to do with global warming. A mathematical model was built describing these cycles of sudden Arctic warming. It turns out they are influenced by the nature of atmospheric circulation that in turn is defined by the North-Atlantic fluctuation index and the speed of Earth rotation (which is known to be not constant but rather have some spread). The higher is the planet rotation speed, the warmer it gets at high latitudes because of increased transfer of warm waters from the equator to northern latitudes. By the way, the western transfer, which is familiar to us by now, also intensifies. For Europe and Arctic, this means increased amount of precipitation and elevated temperatures in all seasons except fall. But let's not delve into particulars, but rather continue discussion on climate reconstruction . . .

Generally, as simple and pleasant it is to reconstruct climate in Europe, as hard and unpleasant it is to do it in gloomy Russia. Of course, Russia had its own kindly lunatics. There even was an amateur meteorologist, a medical doctor by the name Lerche, like that Icelandic doctor, who conducted meteorological measurements until his death. By the way, this Russian weather calculus fanatic was German. This was not surprising—until the eighteenth century, all Russian scientists were German. But something else is interesting—the first scientist of Russian origin appearing in the eighteenth century, Mikhail Lomonosov, was in fact not even a scientist, because he did nothing for science. But then, he was a skillful administrator of science and able to get money for it. However, that money was not always used for the benefit of the Russian state.

Lomonosov brought mosaic from Europe and decided to master the production of similar mosaic at home, for which he managed to squeeze a large budgetary loan that went nobody knows where. The result is well known: no money, no mosaic. Besides money, Lomonosov liked fantastic projects. Once, a crazy idea got into his head—to reach India and China . . . via the Arctic Ocean. For some reason, Mikhail thought this ocean was ice-free to the north of the eightieth parallel. It's hard to say how such an idea can generally come to one's head, but it did! And the influence of the first Russian "scientist" at tsar's court was so great that he easily got the moneys for two expeditions from Catherine who has just ascended the throne.

Those expeditions ended, naturally, in triumphant failure. The plan was to reach the Pacific Ocean, going round Spitsbergen, Greenland, and North America. An amazing project . . . While the Northeast Passage can still be at times ice-free for passage of ships, the way round North America is impassable even for icebreakers up to now. The first expedition under the command of Chichagov bypassed Spitsbergen in summer 1765 and came to rest against heavy multiyear ice. Next year, everything was repeated. And do you know who caught hell for the expedition failure? Chichagov, of course! He received a formal blowup at the Admiralty Board.

As is customary in Russia, the expeditions were enveloped with a veil of extreme secrecy. The expedition dispatch was kept secret even from the Senate and became known only after the ships were put to sea from the port of Archangel. By the way, Bolsheviks did not invent their love for secrecy—they borrowed it from there, the dark tsarist past. I have no clue as to why Russians were and are in such love for secrecy. But they do love it awfully! Here's another small illustration to that.

In the early seventeenth century, a Dutch merchant, Isaac Massa, spent three decades living in Moscow. One of his Russian acquaintances with access to the court gave the Dutchman an amazing map of Russian Arctic. The map showed the Novaya Zemlya (*Russian for "new land"—Translator's note*) with the strait, the Yenisei River mouth, several islands in the Kara Sea, the western part of the Taymyr Peninsula, and the northern tip of the Yamal Peninsula. The official thought up to now is that all this was unknown then and that first pioneers came to those remote places much later, in the late eighteenth to early nineteenth centuries. Upon returning home, the enterprising Dutchman published the unique map accompanied by the following comment: "The person who gave me this map took a severe risk with a possibility of being beheaded, because the Russian nation is extremely distrustful and doesn't like when the country's secrets are uncovered."

All this is meant to point out that the naval archives where logbooks are stored, including from Chichagov expeditions, which today are of interest only to historians and climatologists, are still practically out of bounds to scientists. On top of everything else, Russia State Navy Archive, the depository of invaluable documents, has closed its doors this year indefinitely. So this is what kind of daredevils we are—we'd rather die than tell a state secret! We won't even disclose it to ourselves under torture. Life is just a penny worth! And on a grand scale of things, all our national secrets are worth the same . . .

Well, enough self-flagellation. Let's better go to China! It's nice in China. People keep track of the weather there for thousands of years. The first discovered texts relate to 2187 BC. It would have been very easy to create climate reconstruction for China, if only in AD 212 one Chinese idiot—namely, Shih Huang Ti, the first emperor of Chin dynasty—were not to issue an order to destroy all books in the nation. He was a true Confucian, that's why he ordered the way he did: according to Confucius, the ruler must keep the people in ignorance. So, the emperor tried his best.

Religion is a scary thing! If science is the torch of intellect, religion is its extinguisher. Long after Ch'in Shih Huang Ti, another similarly sincere religious ruler—caliph Omar who captured Alexandria in 642—ordered in the name of Allah to burn down the Alexandria library that kept scrolls collected for about one thousand years. Here are Omar's famous words that served as a signal to the wildest crime: "If books contain things contradicting the words of the prophet, then they are harmful and should be destroyed; and if they confirm the words of the prophet, then they are simply useless." The invaluable scrolls were burned for a week, heating up Alexandria bathhouses where Arab conquerors relaxed themselves.

. . . Eh, he should have wasted one hundred thousand people instead of burning down Alexandria library! . . .

Only thanks to a miracle, some Chinese texts of pre-Chin times have reached us—like, for instance, the famous Bamboo Book of Annals that had been buried in the tomb of one of local monarchs. Confucians did not want to disturb the remains of the deceased or maybe they just forgot about the book, and that's why we have an opportunity today to enjoy the weather information for one thousand years before the Chin dynasty—from the mid-second through the mid-first millennium BC. From that time on, the number of sources increases.

Hundreds of thousands of climate information documents preserved in the course of the entire history of China were catalogued and processed in the 1920s by a Chinese researcher Chu Ko-jong whose works were recognized as classic. However, up until the 1980s, China published these works only in Chinese, which made them practically incomprehensible to the world scientific community that knows zilch in Chinese. But fortunately, since the late 1980s, these documents started being published in international journals.

I hope the reader now understands quite well the pluses of historical climate reconstruction. However, the minuses are that this reconstruction is not possible in all places but rather only in those that were inhabited by relatively civilized people. The reconstruction based on historical documents is not possible for either Africa, or Australia, or America. Or rather, it is possible for North America, but only from the time when white people came there. Because those Indians are . . . savages . . .

Speleothem Method

This method deals with studying cave deposits, as you perhaps might have guessed. It was found that stalagmites (the ones that grow from floor to ceiling) contain annual rings just like trees! These rings are formed because drops of water from rain or melting snow are dropping unevenly. There are more drops during the rainy season or spring, while fewer in winter. And by the way, these drops contain a fixed amount of oxygen isotopes ^{18}O and ^{16}O, which as we remember linearly correlates with temperature.

So, it's an ideal paleothermometer! The layers can be easily counted. Materials are easily dated using uranium-thorium method. There are about a dozen speleothem laboratories in the world. For example, there is no dendrochronology in Israel at all because there are practically no trees there—either desert or small crooked bushes. But they do have caves there! And they do have paleoclimatology.

Instrumental Observations

As it's easily understood, this is direct weather observation: barometers, thermometers, hygrometers . . .

Thermometers were known since the early seventeenth century. They were invented in Florence, Italy. But first thermometers were bad because they produced irreproducible readings: primitive construction precluded from making accurate measurements, causing all thermometers to produce different readings. So if you somehow come across temperature measurements from the seventeenth and early eighteenth centuries, please perceive them without fanaticism.

The first lot of reproducible thermometers (with the same readings for the entire lot) was manufactured by Gabriel Daniel Fahrenheit in 1727. That year is considered the birth date of precise thermometry. Fahrenheit was a strange person—for some reason, he assigned 100 degrees to the temperature of a sick patient with high fever (37.8°C). It's a riddle as to where he found a high-fever patient every time he had to adjust the instrument. But the temperature variety did not end with this Fahrenheit oddity. Several more temperature scales appeared in the eighteenth century. Manufacturers of thermometers, just like first manufacturers of video, could not agree initially on single formats and standards—they all aimed at making their own. There was Celsius scale, Reaumur scale, and even Delille scale that was turned "upside down" with 150 degrees for ice melting temperature and zero for water boiling temperature. In England, they used the idiotic Fahrenheit scale that was later moved smoothly to America. The French who were freezing near Moscow used the Reaumur scale and were scared seeing –20 degrees on it corresponding to –25 on Celsius scale (-13°F). Reaumur was by the way a bloke too, not without oddities: for some reason, he marked the water boiling temperature by the number 80 and the water freezing by zero, which caused the "thread pitch" to be different from Celsius, albeit somewhat close.

By the way, Kutuzov viewed Reaumur the same way as Napoleon did: Russia used the Reaumur scale then and moved to Celsius by the end of the nineteenth century—along with all of Europe, when the single metric standard was introduced. England officially changed to metric system only hundred years later—in 1970, but they in fact still measure everything in pounds, yards, Fahrenheit, and other moronities. It was only recently that unified Europe forced those snobs to come to their senses. But even today, Brits understand pounds and Fahrenheit better than kilograms and Celsius. And the difference between Celsius and Fahrenheit can be quite substantial. If a Russian at 20 degrees starts preparing to go to the beach, a Brit hearing it's 20, knows that it's time to get the shovel and dig out his car from under the snow. But sooner or later, the Euro zone will win and they will get used to it in England—where else can they otherwise go?

And only the United States stubbornly resists the advance of civilization. That's why their temperature up to this day is that of a high-fever patient.

*   *   *

So in general, you saw that paleoclimatologists have many climate reconstruction methods, some of which are amazingly accurate. Some scientists specialize in dendrochronology, some in limnology, and others in something else. As to Klimenko, he is one of those few who are involved in system analysis. In just fifteen to twenty years, he accomplished a simple scientific feat that went unnoticed by the general public—he brought all the known paleoclimatic data together into one general picture and reconstructed climate of Holocene in the Northern Hemisphere. For that, we are saying our big human "thank you" to Klimenko. On this note, we're done with the boring part of the book and turning to a livelier one.

CHAPTER 3

Beginning of Beginnings

Civilization's history began approximately five thousand years ago. However, even before this momentous date, people existed perfectly well for thousands of years—built cities like Jericho and Catalhoyuk, sowed and plowed, and drank beer. It was this era, the prehistoric era, that remained in humankind's memory as the golden age, for it was warm and comfortable on the planet then. And this bliss lasted about 4,500 years (from 10,000 to 5,500 years ago).

It was blissful time. But it was unnoticeable on the scale of historic events, because we already know that humankind makes global breakthroughs only at those times when humans suffer. And when humans are doing fine—why bustle? But life was not easy before the golden age, thus creative breakthroughs did happen.

Seventy to seventy-five thousand years ago, the entire equatorial Africa suffered a severe drought. At least that's the picture painted after drilling the bottoms of African lakes Malawi, Tanganyika, and Bosumtwi. The Lake Malawi (it extends today 550 kilometers [340 miles] long and 700 meters [2,300 feet] deep) had all but dried up in that era, becoming a scanty chain of lakelets less than 10 kilometers (6 miles) total length and less than 200 meters (650 feet) deep. And the Lake Bosumtwi lost all its water completely. It's quite possible that it was this climate catastrophe that forced our ancestors to start moving and inhabiting the world. This hypothesis is especially easily accepted considering that humans are water monkeys (for details, see my book *Monkey Upgrade*), preferring lake shoals.

Fifteen to twenty thousand years ago, the humankind was experiencing the peak of glaciation—the worst cooling in its history. It was in these hard times when the humankind achieved grandiose successes—humans have once and for all settled down on all continents, mastered fire, revolutionized hunting by inventing long-range weapon (bow), invented arts, and mastered well-developed speech. The times were tough, but fruitful. They were replaced by calm and fruitless era of the golden age that lasted through the end of the fourth millennium BC.

And then the history began . . .

PART 3

Average Hospital Temperature

Let's imagine the weal of the people of Ancient Egypt. They sow wheat, pasture cattle, make beer, and otherwise live well, what else do they need? But no, they jib and build giant Pyramids—sweat, labor, dust, stones, knock, heat . . . Why couldn't they just lie in the shade?

Mikhail Veller

"Have you been to Namibia?"

Klimenko has the ability to ask ridiculous questions, after all. If one thousand Muscovites were asked whether they had been to Namibia, I could bet 99 to 1 that none of them had been to Namibia!

It's one of those questions that need no answer. No, I did not fly in an air balloon. I don't know when we had the coldest (as well as the warmest) winters (and summers) in the last five hundred (two thousand) years. And I have not been to Namibia.

"What would I be doing there?"

"Oh well! It's a very interesting country."

Klimenko speaks about all countries that they are interesting. Klimenko is an experienced traveler. The person studying the planet. There are much fewer countries in the world not visited by him than the ones he visited. The places where Klimenko's foot has never stepped could be counted by fingers on both hands. Once, when I was at home just drinking tea, the phone rang and Klimenko's voice asked:

"Would you like to join me on a trip to Greenland, Iceland, and Faroe Islands? Just take a jacket along. It's cold there now. And sneakers—to have something on when jumping on wet rocks."

A jacket? I looked out the window. The poplar down was flying, the heat was near thirty (mid-eighties in degrees Fahrenheit), and I was planning to go swimming in a warm sea. But I don't like the north. And a business trip to the Chukchi Peninsula in 2000 just convinced me in that attitude. Yet, I might perhaps fly to Africa. But why specifically to Namibia? And then, they are probably shooting over there . . .

"No, on the contrary! In Namibia, everything is quiet. Namibia generally is an affiliate of Germany—former German colony. And with

Germans—you just can't take it easy. They still have many German schools there and Namibian school diplomas are accepted in universities and colleges of Germany on par with the German ones. The capital of Namibia has plenty of bars, and the walls of those bars bear photographs of the dear fatherland. There are posters "Munich-Bavaria" across the street, and German beer is sold everywhere. Streets are named accordingly—Bismarck-strasse, Goering-strasse . . ."

"That same Goering?"

"Almost. This is Hermann Goering's father. Hermann Goering was born in Africa and graduated from high school there, but when World War I started, he went to Europe and volunteered to go to the front, becoming an ace fighter pilot. And Goering's father was a governor in Africa, and a good one he was, which is why a street was named after him. Population of Namibia is only about one million, while the country is vast, and so there's plenty of room for everyone. Back when Germans were there, they lambasted aboriginals heavily. In a cathedral in the country's capital, Windhoek, hangs a long list of Germans perished when quelling the seventh revolt (!) of Herero, one of many aboriginal peoples. And hanging somewhere are probably the lists of Germans perished in six earlier revolts, although I saw no lists of perished Herero—they just wouldn't have enough walls for them. Three quarters of the Herero nation was destroyed. But today it's peaceful and quiet in Namibia."

"Good for them. True civilizers. But why did you all of a sudden mention Namibia?"

"Well, because its population is very interesting—simply a preserve of races. And by the way, it's not the Negroes that are aboriginals of that area, but Khoisan group peoples—Bushmen, Khoikhoi (Hottentots), and Mountain Damara. They are not even black, but closer to Chinese—their skin color is yellowish and Negroid features are feebly marked and introduced by latest 'extraneous influence' of Negroes."

"Hottentots . . . I heard something about them."

"Of course, you have! The first colonists there were the Dutch, who married exclusively Khoikhoi girls. The reason was that Khoikhoi women had an exceptional feature—huge round butts. So there's this whole Métis people living today in Namibia—it's called Rehoboth bastards, and they still live in strict accordance with the Dutch way of life of the seventeenth century."

"Big-butted! I remember! If that butt were slapped in the morning, it would still be shaking like jelly in the evening. With the form of their butts, they greatly attracted Dutchmen who had a weakness for cellulite."

"Yeah, the Dutch liked that stuff very much—just look at the women of Rubens or Brueghel. The native people of this area were so much nonblack that during apartheid times they were not even considered Negroes. There were four groups of people then—white, colored, Asians, and blacks. The 'worst' category was blacks. The 'best,' as you realize, was white. And Khoikhoi, Damara, and their descendants of mixed marriages were considered colored."

"And I always thought that Africa is the birthplace of Negroes."

"That's where I'm leading! Negroes appeared in Africa, south of the equator only 2,500 years ago—during the times of Herodotus, Aeschylus, and Sophocles . . . They managed to cut their way through from the north. Cutting their way through the tropical forest became only possible after the invention of iron and its gradual penetration into Africa. It was the Axial Age—the time of great migration of peoples, invention of iron, appearance of world religions. The time of big global cooling leading to movements of peoples and peoples' minds. But there were other cooling periods before the Axial cooling. Like the one that led to the beginning of human civilization. It happened more than five thousand years ago.

CHAPTER 1

Mythology of Olden Era

In the late fourth millennium BC, an event happened, from which historians start the count of history of not humankind but civilization. In about 3100 BC (plus or minus 150 years), Pharaoh Menes united North and South kingdoms (Lower and Upper Egypt), thus marking the beginning of the so-called Old Kingdom.

Let's remember this date—3100 BC. Because it was at that time, approximately five thousand years ago, three greatest civilizations suddenly flared up on the populated planet like three torches, simultaneously and independently of one another—Egyptian, Mesopotamian, and Indo-Harappan.

Public at large is familiar with Egypt; it is less familiar with Mesopotamia, although everyone heard, of course, of Babylon and Sumer. But the Indo-Harappan civilization is known just to a narrow circle of population and very "narrowly"—even specialists know all but nothing about the history of this greatest culture: it is mute because Indo-Harappan written language isn't yet decoded, unlike Egypt hieroglyphics or Babylon cuneiform.

Egypt never left the mind of progressive humankind. It was always near, close by. It was conquered by armies of Alexander the Great, Romans, Arabs, Turks, Brits, Napoleon . . . And each European conqueror brought along a crowd of scientists who immediately began studying and recording everything.

Humankind discovered for itself Mesopotamia only in the nineteenth century when the weakened Osman Empire started slowly allowing first scientists into its territory—Germans obviously, because they were

—

111

allies. As soon as the iron curtain fell, crowds of high-class German archeologists poured into the lands of Turkish Empire. German scientists had a grandiose task before them, to read the Bible as a history document, hoping to find all the cities mentioned in the Book of Books. They did!

. . . School textbooks of the nineteenth century have no mention of Sumer, Babylon, or Mesopotamia, because at that time, Germans have not yet worked with shovels and brushes on the Osman lands. But a lot was known about Egypt. Even modern textbooks are overfull of information on Egypt. They have much less on Sumer and not a word on Indo-Harappan culture. The latter was discovered only in 1920s, and there's simply nothing to write about it yet because of its muteness. But discovery of this culture shocked historians: suddenly out of nowhere, the greatest ancient civilization appeared, and nothing was known about it. It's the same as finding an extra car in one's garage or an extra armchair in one's home.

In the Indus river valley in the territory of today's Pakistan and northwestern India, remnants of more than one hundred cities were found whose total population exceeded one million. It was discovered that those cities had water supply and sewerage systems, social buildings, vast granaries, residential and worship structures, and most importantly, huge number of written sources. Unfortunately, they are still not decoded, preventing thus to learn more about this civilization. The only thing known is that this civilization lasted for over 1,500 years (longer than Rome) and that its cities were not surrounded by fortress walls, as were the cities of the era of the Roman Empire's highest might, similar to the modern cities. An extremely revealing fact, isn't it?

People lived all over the planet then, so why do the specialists select these three cultures? It is because—among the overall populated space, they stand out like three peaks on a plain. They were high cultures! Historians call high cultures those that have cities, division of labor, centralized power, legislation, and written language.

Up to the very end of the fourth millennium BC, there were practically no high cultures on the planet. And suddenly a whole three flared up, almost at once! And these civilizations were totally isolated from one another and during their early years knew nothing of one another's existence. It's a strange chronological coincidence. So what made them flare up?

Climate worsening.

The rise of all three civilizations coincides exactly with the era of global cooling. The cold has crystallized civilization. Or rather, the main role was played here not as much by cooling, but by its induced drying up of climate, leading to reduced crop yield and forcing people to combine their efforts, that is, establish a state.

When life roughly takes you by the throat, you have to come unscrewed—think, invent, try. In order to increase crop yield that was catastrophically falling, irrigation installations were needed. All peasants needed irrigation canals, but they couldn't construct a canal acting alone. Therefore, efforts of all peasants had to be combined. But how to force individual peasants to work for the common good? Only by force. Thus, an enforcement apparatus appears—the state. In order to carry out large projects, such as irrigation systems, a strong mighty centralized power is needed. And the existence of such power presupposes a developed tax system; or else the power would have nothing with what to feed its enforcement apparatus. And tax collection requires well-developed written language and special tax services. And thus, you get civilization, or complication of social structure for the purpose of survival.

The vast territories of America were still inhabited at that time by various disconnected tribes, whose main occupations were hunting and collecting, while farming was extremely primitive. But, the climate crisis at the end of the fourth millennium BC apparently had its effect upon even these heavenly places: with the drop in heat and moisture, appropriating economy proved ineffective. Not incidentally, this was the time when a whole range of new crops was grown in Mesoamerica and Andes area—cotton, batata, yam, and capsicum—although before that, for thousands of years, people were content with only corn and beans. And this was the time when the first earthenware on the territory of America appeared in the Andes area.

So how much has the climate worsened?

The average global temperature dropped 1°C (2°F). Before this fateful cooling, the Earth looked entirely different. There was no sign whatsoever of any Sahara desert. In place of today's boundless sands was no less boundless savanna, and the main occupation of its inhabitants was cattle raising and fishing (in numerous lakes and rivers) and to a lesser extent farming. The Great Nile was not the same river as seen on today's maps. Today, the Nile's last two right-hand tributaries, the Blue Nile and the Atbara, flow on the territory of modern-day Sudan. And the Nile has no left-hand tributaries because the Sahara desert lies to its left. However, in the times that we're talking about, the Nile had numerous tributaries particularly on its left side.

—

The signs of these mighty rivers were discovered about thirty years ago with the use of space and aerial photography. Naturally, after such a discovery, scientific expeditions started setting out to the desert and drilling former riverbeds of former rivers. Remnants of the greatest variety of tropical fauna were found—like crocodiles, hippopotamuses, ostriches. The dating showed that the most recent remnants are four thousand years old. In other words, Sahara was not at all a desert even seven hundred years after construction of the Pyramids!

The sands, once having started their attack against people, were advancing slowly. So ancient Egyptians up through the Middle Kingdom observed the landscape around them that was entirely different from the dismal landscape seen today by Russians vacationing in Hurghada. Herodotus left us the most interesting evidence of a travel undertaken by Nasamons (the people living then on the lands of modern-day Libya) into the depths of Sahara. The travelers discovered there "vast marshes, then they arrived in a city whose all citizens had black skin, and a big river was flowing past that city from west to east . . ."

Lush vegetation and populous cities—this is how the Sahara looked! Precipitation amount exceeded current level three to four times. Today, Egypt has 50-75 mm (2-3") annually, while it had 300-400 mm (12-16") then. The same could be said of Mesopotamia and the Indus valley. People were doing very well. In fact so well, that they had no need in a superstructure in the form of a state. But at the end of the fourth millennium BC, as we already know, the cold and the drought crystallized three high civilizations.

After some time, the climate changed again—toward the better: it became substantially warmer on the planet with more precipitation. The powerful Egyptian, Mesopotamian, and Indo-Harappan cultures continued to flourish. But then, as it usually happens with climate, it fluctuated again—not toward the better.

. . . These climate muddles should not be shocking—just as daily and weekly temperature fluctuations are not surprising, so neither should be constant climate fluctuations. Climate is the weather of the millennium . . .

In about 1500 BC, the average hemisphere temperature again dropped by almost one degree (2°F). This climate change provoked some dramatic events in the world. From somewhere in the north, all of a sudden numerous barbarian tribes came—the Aryans, whose onslaught crushed

—

the great Indo-Harappan civilization. The Aryans were led by leaders unknown to us, like Attila, Genghis Khan, and Alexander the Great.

But what forced the savage nomads out of Central Asia where they lived carefree until then? Famine. Climate worsening in the steppes of Central Asia created deficit of heat and water. Consequently, people had not enough food. And food shortage is none other than population surplus, isn't it? And this surplus population, in order not to be snuffed, took off and went where their fancy took them, as they thought . . . But according to the climate map, they went in fact in the direction of moisture gradient. They came to the lands of Indo-Harappan civilization. And wiped it out. They simply had no other choice. Back home, the Aryans faced a certain dull death from famine; ahead, it could be an honorable death in battle or hope for a win in a climate lottery.

These savage Aryans, as you might know perhaps, are our direct ancestors. They invaded the "civilized space" from Central Asia where they had arrived earlier from Southeast Europe that was covered at that time with a wave of severe cold and drought. They invaded and brought along their strange language of so-called Indo-European group. I shall not bore you here by listing all the languages of the Indo-European group. Instead, I'd rather give you a picture of the world, showing all the countries using the languages of this group (fig. 3). After all, a picture is worth a thousand words.

Fig. 3. Distribution of languages on the globe:
■—Indo-European language group

This map is a remote result of climate events happening three thousand five hundred years ago (about 1500 BC).

Our planet history had numerous mass migrations of people, similar to the Aryan. And all of them (including the so-called Great Migration of Peoples and crusades) were caused only and exclusively by the "bad climate." Indeed, what would make people leave their homes and start conquering other lands while risking their lives, if life in their own land were good and plentiful? They lived well there for hundreds years, and all of a sudden, as if a rein got under the tail—they took off to conquer foreign lands. Such sharp forms of competitive social struggle like wars are not born as consequence of good life.

. . . The fall of Indo-Harappan civilization is reflected in the most ancient written sources that humankind has—the Ancient Indian Veda scriptures written in thirteenth century BC, just two hundred years after the invasion, practically hot on the heels of the events. The Vedas tell about destruction of cities from the invasion of northern barbarians. And it was since those times, a new god appeared in the oldest Indian religion (Brahmanism), the City Destructor . . .

Those who do read books attentively might have a reasonable question: if cooling facilitates the rise of human spirit, inventiveness, establishing empires and civilizations, then why did the cooling of 1500 BC kill the Indo-Harappan culture rather than facilitate its development?

It was because at times of climate worsening the victors are always those who are in a worse situation: they have nowhere to retreat. This time, the Aryans were the ones who had no way back. They had to either die or find a new place to live, even while taking it away from others. And that's what happened. Desperation is the most direct way toward victory.

. . . It's like a locust swarm . . .

And what was going on meanwhile in the other two centers of the world culture? How were Egypt and Mesopotamia doing?

The Old Kingdom, born at the dawn of civilization, lasted successfully until about twenty-third century BC. It marked its existence in the book of eternity by building great pyramids, after which it collapsed. The collapse of the Old Kingdom was always a puzzle to historians. They put the blame for it on climate worsening that the social system allegedly could not endure. In fact, historians knew that at the end of the third millennium BC, Egypt suffered severe drying up, causing huge crop failure lasting two to three years. This natural catastrophe has a witness in the form of one of the most significant written sources of that era—the famous Palermo stone that marks the signs of terrible crop failures of that period. The drought was so severe that the Nile could be waded across—an event so rare that it happened only several times in the last millennia. (For your information, the Nile average width in Egypt exceeds 1 kilometer or 0.6 mile.) Historians believe all this shook the pharaoh's power, since the pharaoh was to peasants a live god responsible for all, including the crop and the Nile overflows.

However, refined paleoclimatic picture disproved this hypothesis. The Old Kingdom began cracking and collapsing in the era of good climate, but the peak of climate troubles came one and a half centuries *later*, in the twenty-first century BC. The terrible period of internecine feuds in Egypt lasted from 2250 through 2070 BC and ended when a new pharaoh came, Mentuhotep, who became the sole sovereign ruler of Egypt and founder of a centralized state. Historians call the era that started with accession of Mentuhotep the era of Middle Kingdom. And we should not be surprised anymore that the rise of the Middle Kingdom happened at the very peak of the climate worsening.

So, it's the twenty-third century BC, the start of gradual decline of VI dynasty of Egypt pharaohs. The worsening of climate has hardly even begun then, and it was still about two hundred years before the climate catastrophe. So why did the decline begin?

When life is good, crops abundant, and energy resources plentiful—then nuts and bolts of centralization loosen involuntarily, and power and influence move naturally to provinces, because the multiplied offspring of local bureaucrats need some warm places and nice food. Resources are needed for both power and finances. The center pays no special attention to local princelings, now corpulent and impudent, because the collected taxes suffice for everyone's comfortable life. And now this state has everything ready for its demise. It hangs by a thread and by old glue of past habits and laziness to any changes. But a smallest shock leads to the collapse of loosened blocks. And so, when the climate curve starts its swift dive from blessed temperatures to a famine era, the country collapses, of course. The center had no more levers of influence and the local princelings had no desire to submit to the weakened center. Unrest . . . Aggravated by starting spell of climate troubles.

Paleoclimatologists note that between 2500 and 2200 BC, water level of the lake system Zway-Shalla on the Ethiopian high plateau has dropped 100 meters (330 feet), causing the once large fresh-water lake to be transformed into a cascade of small salt lakelets. And studying fossil flora on territories of modern-day Burundi, Uganda, and Rwanda revealed that in the period from 2200 through 2000 BC forest was pressed by brush and grass. Drought . . .

Advance of drought is seen also by the fact that royal tombs were coming closer to the Nile coasts. The point is that each royal tomb was built in such a way that the Nile overflow waters would touch the lower steps of the smaller temple. A royal tomb of any self-respected late king was a set of structures consisting of a required "gentleman's set": a burial vault pyramid with an upper temple at its foot and a pathway from it to the lower temple. The steps of this temple were supposed to be licked by the waters of the overflowing Nile. Well, archeologists have long noted an oddity: with each next deceased king in the twenty-fifth to twenty-third centuries BC, the temple complexes came closer and closer to the Nile. This means the Nile overflows became weaker and weaker.

Crops were falling with each year. Egyptian written texts of that era, which survived the times, bristle with information about grain crop failures, droughts, and cannibalism. If you're not too lazy to look at the graph below (fig. 4), you'll see that the average hemisphere temperature drop then was 0.5°C (1°F). This temperature drop transformed Egypt from the mode of "Mediterranean granary" into the mode of "cannibalism." As was customary there in such situations, peasants ate their own children and killed occasional passers-by.

The global cooling was caused by a set of reasons. The long-cycle periods of solar activity fluctuations were superimposed on one another: 2400 year, 1100 year, 850 year, and 640 year. This caused aggregate decrease of solar activity, causing decrease of energy reaching the Earth's surface, causing in turn cannibalism. Of course, the half-degree (1°F) cooling did not deliver the fatal blow to Egyptians by itself. It's just that when temperature drops, precipitation regime changes in tropics and subtropics (which is related to reduction in the influence zone of summer monsoons). And reduced precipitation leads to reduction in rain forest area, increase in desert area, drop in water levels of Rift Valley lakes, reduction in river flows, and droughts . . . That's what makes global cooling so dangerous for North Africa. And as you know, Egypt is in North Africa.

But during global warming, everything happens in the exact opposite way. And the interesting thing is that a slight global warming causes both winter and summer temperatures in North Africa to *drop*! This happens because warming leads to higher precipitation in the region and consequently to higher evaporation. And the process of evaporation of liquids, as known, is associated with loss of energy. Everyone knows that a wet cloth is cooler than a dry one and the wet thermometer of a psychrometer always shows lower temperature than the dry one (see high-school course of physics). Thus, global warming is very beneficial to North Africans—it's both good for crops and not too hot . . .

But let's get back to the very peak of cooling, when the iron fist of Mentuhotep brings Egypt into a unified state crystal. The point of unification is that it's easier to fight the elements together. Under the keen leadership of their new father, peasants are building irrigation canals in order to provide a more rational use of the scarce water. It was under the rule of Mentuhotep that the unified irrigation system was built in the country (similar to the Russian unified energy system owned by Russia joint-stock company Unified Energy System). The pharaoh introduced the strictest accounting and control of crops.

Worsening living conditions caused a fundamental administrative, technological, and social restructuring in Egypt. For example, the collapsed Old Kingdom was essentially a socialist state—work in stone quarries, on pyramid construction, in fields was done by special "detached force" teams receiving a certain reimbursement for their labor from bureaucrats. It would be a great stretch to call these people peasants. But as Egypt rigidly pulled together into the Middle Kingdom at the peak of the climate minimum, land cultivation became organized more progressively, or more "capitalistically"—the state had to renounce its collective farms and transfer the land to peasants' hands. They were, of course, not "individual peasants" but the ones assigned to the land and fulfilling their labor quotas, but even serfdom was a big step forward compared to labor camps.

An interesting point: in the civil unrest, which erupted in Egypt after collapse of the Old Kingdom and led to creation of the Middle Kingdom, the victor was the north. That is, the victor was again the side that was worse off. No, no—climatically, the country's north and south did not differ! But the reduction in water flow forced the citizens of southern Egypt to dig so many individual canals for irrigation of their fields that most of the Nile water was used up "on the way down" and too little left to reach the north. This doubtlessly aggravated the already very bad situation for northerners. It was for this reason the Mentuhotep dynasty came from the north (from Thebes), becoming new rulers of the Middle Kingdom. The one who is worse off becomes the victor.

... Another two hundred years passed, and things became so good in Egypt that healthy herds and rich crops totally eclipsed Egyptians' memory of the period when the Nile resembled the Moskva River and Egyptians ate human meat. The warm weather that settled in the nineteenth to seventeenth centuries BC was the kind unknown to humans. It was the warmest and the most blessed period in history of civilization.

While the Nile water-flow during the wretched downfall times was 90 km³ (22 cubic miles) a year, now it was never less than 160 km³ (38 cubic miles). Precipitation increased two-three times. The full-flowing fat Nile crawled lazily toward the Mediterranean Sea, and its overflows became so full that people even had to fight them! For the first time in humankind history, an attempt was made to control natural elements, specifically—the Nile overflows. It was an amazing project! Egyptians dug a canal between the Nile and Faiyum Oasis located 80 kilometers (50 miles) west of the Nile in lowland that the canal helped to fill with the Nile water. This resulted in creating an artificial reservoir—the Lake Moeris, on whose shores a new capital was built—the city of Kahun. This water reservoir was supposed to serve as a buffer water-catching basin—to catch excessive water coming through dams during catastrophic Nile overflows and to use this water later for irrigation.

Here's the point. The drying up and small overflow of Nile are bad because they bring about crop failure and famine. But excessive overflows are also bad because high floods wash away villages and destroy urban structures. So, Egyptians decided to slightly improve the nature. Do you understand the scope of the idea, its organization, and implementation? And all this was accomplished four thousand years ago!

By the way, the new city of Kahun on the shores of an artificial lake was also built in a new fashion, in line with the times. Peter the Great later built his new, northern, capital (St. Petersburg) similarly—linearly, with streets intersecting perpendicularly. Only unlike St. Petersburg, the rectangular Kahun was also oriented along the cardinal points. It was easy for Egyptian architects to do it because the entire city was created according to a single plan, like a single house.

Life became good in Egypt! Written sources of Ancient Egypt tell us that all this abundance, particularly transformation of the Nile from a shallow creek into a super-full-flowing river, happened within one generation. Egyptians were breathing full lungs! Grandiose construction was taking place, Egyptian cities were ornamented with big statues, and medicine achieved great success. For example, one of Kahun papyri describes symptoms of hysterical disorders, as well as their diagnostics and treatment. And it's interesting that the therapy and theory of women's hysterics, developed by Egyptians, stood in medicine up through the early twentieth century AD. The arts of the Middle Kingdom experienced an unusual flight. This is witnessed not just by major works but by minor plastic arts as well—colorful statuettes from the Middle Kingdom era show people in motion: a baker is baking bread, a peasant is plowing, a functionary is counting cattle, or women are bringing offerings to temple . . . And Egyptians made beautiful bright-turquoise faience vases! It would not be shameful even today to place them in a sideboard.

The Middle Kingdom existed successfully for almost four hundred years, and then it was destroyed in 1650 BC by eastern barbarians—the Hyksos. And once again, the Kingdom was destroyed during a period of propitious climate.

An attentive reader can again become indignant here:

"Wait a minute! If climate was propitious, then where did the Hyksos come from? Doesn't your theory tell us that barbarians get on a move only when climate worsens?"

You are right, my attentive reader. I forgot to clarify one thing: climate is a "spotty" phenomenon. Or, as specialists say, "climate fields, and especially fields of humidification, strike us with their irregularity . . ." While all was good and even wonderful in Egypt, Middle East and West Asia had significant drying up of steppes. Famine ousted the local population that started roaming in the direction of the humidification gradient in search of more fecund land. And this land was found—in Egypt. And that was the end of the Middle Kingdom.

It was a good civilization! But it collapsed at approximately the same time as the Indo-Harappan culture did. Everything collapsed quickly and suddenly—as a notable German historian wrote, "as if the light suddenly went off." But the Hyksos, who put out the torch of civilization, held power in Egypt for just about hundred years. During that time, several pharaohs of barbarian origin replaced one another on the throne of Egypt. And then, the same climate worsening, which ousted the Hyksos from their homes, reached Egypt. And as soon as it happened, Egypt livened up, came together, and quickly kicked the butts of barbarian pharaohs. A local Egyptian resolute chap appeared and unsentimentally wrung Hyksos' necks. His name was Ahmose. He became the first pharaoh of the New Kingdom.

This was the classic case of crystallization of a state! Ahmose structured the country rigidly and appointed two deputies responsible for the country's north and south respectively. All local functionaries were obliged to send regular reports on the state of their respective territories to Ahmose deputies. The military machine was reformed with army converting to garrison system. All in all, the power vertical was successfully restored.

The climate worsening was so significant that the quickly restored country began demonstrating imperial features itself, or in other words, embarked on the path of expansion. Egypt never overstepped its boundaries before—neither during the Old Kingdom times, nor during the blessed era of the Middle Kingdom. And now it became a classic empire—that is, it began to conquer and annex its neighbors.

Ahmose conquered Palestine and Syria. The cause of Ahmose was continued by Pharaoh Thutmose who replaced Ahmose at this post and moved the troops west, overriding part of Libya. At the end of his reign, Thutmose made a trip south, to Nubia, and extended the boundaries of Egypt all the way to the fourth rapids of the Nile (modern-day Sudan). In the north, the Egypt Empire at the times of the New Kingdom was extending up to the borders of modern-day Turkey. So that's how stout and fat the country became!

Incidentally, it should be said in all fairness that the New Kingdom was not the first empire on our planet. Historians believe the first empire was the state of Akkad under Sargon. It came into being about one thousand years before the New Kingdom. The reasons for its rise and collapse still puzzle historians. But not climatologists!

So, Sargon the Great... *Sargon* in Akkadian means "victorious king." (Just don't confuse him with the Assyrian King Sargon who lived 1,500 years after Sargon the Great.) Akkadian Empire of Sargon the Great came into existence at about 2500-2300 BC, just at the time of decline of the Old Kingdom. It was warm and humid then in Egypt, or good, while in Mesopotamia, it was warm and dry, or bad. And we already know that in places where it's bad, a victorious army is born. And that's what happened. A bellicose person appears who beats his enemies right and left and builds the world's first enormous empire extending from the Mediterranean Sea to the Persian Gulf. The empire stretches north to the salt Lake Van (modern-day Turkey) and south to the Arabia Desert.

This was the first experience of empire creation in humankind history. Sargon's Akkadian Empire lasted 250 years and collapsed under attacks of barbarians just as the climate situation on its territory changed for the better—again it became warm, humid, and placid, and people relaxed. And territories of barbarians who routed the Akkadian Empire experienced just the opposite change—it became colder and drier there. And by the way, this was the same time when Egypt was experiencing awful period in climatic sense (the Nile could be waded) and was because of that in the process of coming together into the Middle Kingdom.

A strange thing: it seems that Egypt, Palestine, and West Asia—they all are close to one another. But where climate is concerned, these areas often find themselves in opposite phases not as much by temperature but rather by humidification: when climate improves in Egypt, it worsens in Mesopotamia. And when it worsens in Egypt, conditions improve in Mesopotamia.

With the rise of average global temperature, water evaporation from oceans increase and precipitation also increases accordingly in most areas of the globe. But not everywhere! In Asia Minor, Middle East (without the Syrian Desert, right-bank area of the Euphrates, and Arabian Peninsula), Armenian Highland, and western part of Iranian Highland, amount of precipitation falls sharply due to the northward shift of subtropical high-pressure zone. It happened in the past, and it happens today in the times of global warming. That is, when it's good for everyone—it's bad for the Arabs. Isn't this the reason for their heightened aggressive activities today? . . .

Moreover, the world climate picture is also complicated by the fact that different areas of the globe experience climatic effects that are occasionally not only of different directions, but also asynchronous. In the sense that the temperature drop or rise in a certain region could lag behind the global trend or outstrip it. That is, someone is already suffering while someone else is not yet lifting a finger. The maximum time divergence between global and regional curves could reach a quarter century. And climate changes are felt initially in West Asia and only later reach Europe.

It would be interesting to look at the climate swing in Egypt to West Asia. History of armed confrontations between Egypt and West Asia

(Hittites and Assyrians) follows precisely the fluctuations of this swing: the victor in these confrontations was always the one who was worse off in the climate aspect.

The same is observed in relationships of Rome-Byzantium with the Persian-Parthian state. It's known that armed conflicts between these two giants lasted for almost one thousand years. And here again, we see the effect of the climate swing. The point is that during global cooling, it becomes colder and drier over most of the Roman Empire, while it becomes cooler and more humid in Parthia. For Rome, the temperature alternation of warm and cold winters is much more important than the amount of precipitation. While for Parthia, fluctuation in the amount of precipitation is more important than temperature fluctuations. In this sense, Rome and Parthia were always in "antiphase." That is, global cooling provides less propitious conditions for Romans and more propitious for Parthians. And global warming provides the opposite. And so, in the endless Rome-Parthia wars, the victor was *always* the side that had worse climate conditions. And it was constantly like that during hundreds of years. We'll be discussing in detail the upheavals of this confrontation in the next part of this book because it contains detailed temperature graph for that period.

Chapter 2

History of Temperature

A phrase flashed a bit earlier about relaxation, which was probably not taken by the reader seriously. For that reason, I'll repeat it: "Sargon's Akkadian Empire lasted 250 years and collapsed under attacks of barbarians just as the climate situation on its territory changed for the better—again it became warm, humid, and placid, and people relaxed." This phrase has more sense than joke.

Ancient Greeks, who understood a lot about life, also knew about relaxing effect of climate. The Greek historian Herodotus tells about a notable dialogue that took place between the Persian King Cyrus and his sages. It was the time (late sixth century BC), when the Persian Empire was established, headed by the terrible Cyrus. Life for Persians was tough and hunger-stricken, and they itched to get into an empire. Cyrus easily captured Asia Minor and Greece where climate conditions were better than in his homeland.

So now, some sages were trying to persuade Cyrus to resettle in the captured lands. And wise Cyrus, having brilliantly read the climate card, answered: "In propitious lands people are usually coddled, and one and the same country cannot produce wonderful fruits and raise valiant soldiers."

"Persians then agreed with Cyrus's opinion and renounced their intention. Themselves owning meager land, they preferred to hold sway over other peoples rather than to be slaves on rich plain," writes Herodotus.

. . . Herodotus is a marvel. We shall definitely talk more about him later . . .

And now I humbly ask the readers to exert themselves and take a benevolent look at the picture that I dare to adduce here (fig. 4). It shows the magnified one-half of one scale division that we talked about when viewing the graph of temperature fluctuations for four hundred thousand years (fig. 1).

Zero on the temperature axis is the climate norm for the mid-twentieth century. It means that the graph shows global climate deviation from the twentieth-century norm. Number 1 on the graph signifies the temperature that was reconstructed with the use of paleoclimatology methods. Number 2 is the calculated curve obtained with the use of the global climate mathematical model created in Klimenko laboratory. The coincidence between the two is better than wonderful. Practice proves adequacy of the model that enables to detail the graph by decades and even by single years.

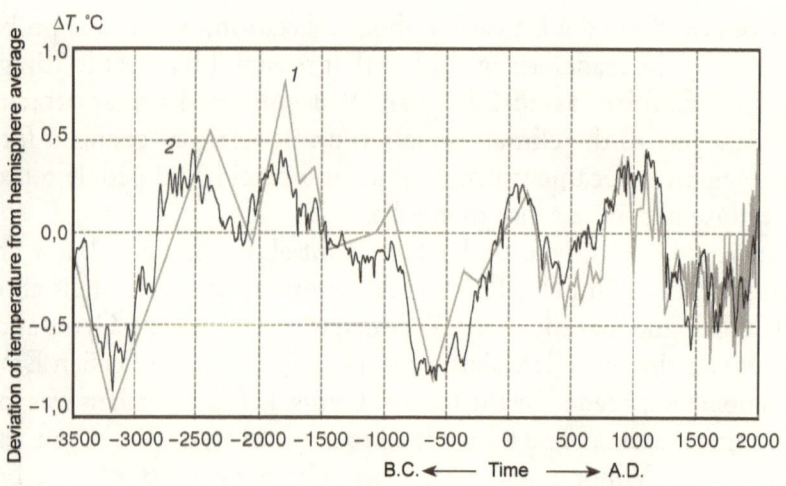

Fig. 4. Temperature history of the Northern
Hemisphere in the last 5,500 years

As you see, the climate cardiogram fluctuated during humankind history within one-degree (2°F) range. It pales in comparison with 8-9-degree (14-16°F) drops during the great glacial eras. However, even occasional half-degree (1°F) drops were enough to shake the foundations.

The temperature drop that gave rise to civilization (3100 BC) was the biggest—the temperature fell then by almost one degree (2°F). It turned cold . . . How big is the temperature drop by one degree if we're talking about annual average temperature for a hemisphere? It's similar to the "average hospital temperature." There's a popular saying about the average hospital patient temperature among Russians, making them believe that this kind of statistics is a stupid and useless thing since it reflects nothing concrete. But that's a mistake. The average hospital temperature can actually reflect something. For example, if the average temperature among clinic patients has risen suddenly by half or full degree (1-2°F), the clinic's senior executive can determine correctly that there's an epidemic in his facility. So if the average global temperature drops by one degree (2°F), it's a sign that regional and seasonal (winter, summer) drops could be 7, 8, or even 10 degrees (12, 14, 18°F), while precipitation changes could reach 300 mm (12") a year. And how much is that? For dry climate regions, it's 200 percent of the norm. In other words, average hemispheric temperature drop by one degree (2°F) causes such drastic climate changes "in the provinces" that they simply can't affect human activities.

Let's go along the graph and see its trends . . . Temperature rise at about 2400-2300 BC (curve 1) relaxed Egypt so much that it saw rumbles of civil unrest lasting through the lower cooling peak that happened in 2100 BC. It's the beginning of the Middle Kingdom. Strengthening of the state during the period of climate worsening.

We remember: when it's nice everywhere—warm and humid, it's bad in the Middle East—drought. That's why in the period of climate paradise, when Egypt being replete began doing stupid things and losing statehood, in the Middle East it was the opposite—very bad and dry. That was when the Akkadian Empire formed there. And it was at that time an event happened that was described in the Bible—Abraham's flight with family from Ur (southern Mesopotamia). It was one of the numerous migrations of peoples known to historians—exodus of Canaanites and Amorites from Mesopotamia to more fecund Palestine valleys.

Next, the second rise of the curve—it's warm, wonderful . . . and collapse of the Middle Kingdom in Egypt at a reading after 1700 BC, under attacks of Hyksos suffering from droughts that brought famine. At the same time, in the Middle East, which is living in the climate antiphase with the whole planet, a fellow by the name Shamshi-Adad founded the Assyrian Kingdom extending from Iranian Highland to central Syria.

Who is Shamshi-Adad? Well—nobody! He just turned up like a devil from a snuffbox. He was the son of the chief of one of the Amorite tribes. There were many like him. Initially he established himself in the center of Northern Mesopotamia, and then he launched a campaign with his band and captured a bunch of small cities in the middle course of the Tigris. One of those cities was Assur that later became the capital. The city gave its name to the newly founded kingdom—Assyria.

As was the custom, Shamshi-Adad began his reign with strengthening the power vertical by dividing his kingdom into several military districts whose commanders reported to him personally. He transformed militia into regular army, and this immediately required establishing tax system. With his new army, Shamshi-Adad the First took a walk north and east where he subdued tribes of Tukrish. Then he went west where he routed the once mighty Yamhad Kingdom in Northern Syria. He captured an important city in southern Syria—the trade center Qatna and went all the way to Lebanon Mountains and the "Great Sea" (Mediterranean). After that, he turned back. He was foolhardy and audacious—and somewhat lucky!

It all ended when Shamshi-Adad set his sights on the pearl of Mesopotamia—Empire of Babylonia. The Babylonian king Sin-Muballit was able to save himself only by paying off high tribute to Shamshi-Adad the First. Those payments to Shamshi-Adad continued for some time by Hammurabi, son of Sin-Muballit. But it lasted not long—when Shamshi-Adad died, his successor was defeated in a sustained war by Hammurabi who conquered his kingdom and then all neighboring kingdoms, becoming the single and undivided ruler of Mesopotamia.

In the era of increasing water shortage, Hammurabi chose a surprising and reliable war tactic—each time he looked for and found the enemy's most important waterway, and blocked it with a dam. And if the enemy still showed resistance, he suddenly opened the dam, causing destructive floods. It's interesting that the contours of Hammurabi empire almost coincided with those of Sargon empire. It means five hundred years later, following the new cycle of drying up in West Asia, the empire was resurrected within the same natural borders!

So let's get back to our curve. The year 1600 BC—against the background of a sudden, sharp, and high-speed cooling, Egypt drives away the Hyksos. After a small and uncertain jump upward, the temperature continues to drop uncontrollably, and Egypt sternly pulls together, creating the New Kingdom that exists happily until the next warming peak in about 1100-1000 BC. And it not just exists but engages, with variable-climate success, in wars against the powerful Hittite state—the victor is always the one whose climate is worse at the time.

But why Egypt is so much affected by every even the slightest cooling? Because in the periods of global cooling, winter and summer temperatures in Egypt rise (the reasons of this anomaly were explained earlier), accompanied by droughts. The living conditions worsen due to crop drop, or in other words, population surplus occurs for which there is not enough food and which burns down not without good in the flames of the next war against the corpulent neighbor. If it's a cooling period—Egyptian armies defeat Hittites. And in periods of warming, when Asia Minor starts drying up, Hittites defeat Egyptians.

During the reign of Rameses II, the two ancient supergiants—the Hittite state and Egypt—clashed in a battle at the city of Kadesh on the Orontes River. It happened at the turn of the fourteenth and thirteenth centuries BC. The climate was then so-so, average, and maybe that's why the grand battle ended in a draw. Generally, in that era when the climate scale fluctuated not knowing which side to take, Hittites and Egyptians fought each other with variable success. Both powers fought long over Syria that was a tidbit because it was a converging point of trade routes connecting Mesopotamia, Asia Minor, Egypt, and Arabia . . . Pharaoh Thutmose conquered Syria and Palestine around 1500 BC, making them Egypt provinces. About a hundred-plus years later, the Hittite ruler Suppiluliuma took Syria away from Egypt. And one hundred years later, the above-mentioned draw-ending battle took place on the Orontes River.

At first, Hittite troops had the upper hand. Having engaged their tactical intelligence, they used an artful maneuver to lure Rameses into a trap and surround pharaoh headquarters. Only thanks to personal courage of the young pharaoh, the course of action was dramatically changed. This is how historians of Ancient Egypt described this dramatic moment: "The Pharaoh tore along and pierced into Hittites. He was alone, nobody with him. And then the Pharaoh looked back. He was surrounded by swarms of chariots with soldiers of the foul land of Hittites and of its many neighboring countries . . . 'There's no prince around, no guide, no infantry commander, or no one else. My army left me for their prey and no one stayed with me . . . So what will happen now, my father Amon? Or would the father not even think about his son? Have I done anything against you? I walked and stood at the will of yours. What are the Asians before you, Amon? They are foul and don't know God . . .'"

. . . It's impossible, of course, to disagree with the pharaoh's latter comment . . .

It seemed this one was totally lost. Pharaoh headquarters is surrounded by superior enemy forces, while most of his own troops have not yet forced the crossing of the Orontes, being bogged down in a fight with the Hittites on the other side of the river. Having nearly just one "commandant platoon" with him, Rameses decided to cut his way out. And he was going to cut in an unexpected direction—toward the river. With a small group of Egyptians, he overran Hittite arrays. This maneuver could have been probably useless for Rameses, but when the Hittite troops burst into the camp left by Rameses they started plundering, and this always ends badly. Hittites' eyes burned with excitement, seeing all the shiny stuff, failing them to notice a small detachment of Egyptians coming to help their pharaoh. The strike was all the more successful as it was sudden, and it turned around the course of the fight. The remaining Hittite troops hid behind the city walls of Kadesh. Rameses could not capture the city, but Hittites were afraid to stick their nose out, too.

...Do admit, you were supporting Egyptians in this fight, weren't you? And how did I guess? But it was the Hittites, who were "ours"! Hittites are our direct ancestors. Studying Hittite libraries where they stored numerous "books"—the earthenware plates with cuneiform—showed that Hittite language belongs to the Indo-European group and its traces are found in many European languages. For example, "eat" is *etsen* in Hittite and *essen* in German; "water" is *wadar* in Hittite, *wasser* in German, and *voda* in Russian ...

After that famous battle at the Orontes River, Hittite-Egypt fighting on the Syrian plains went on non-stop with variable success for about fifteen more years. The fighting ended with signing a treaty of friendship and mutual assistance. In 1272 BC, the Hittite king sent Egypt the text of the peace treaty engraved on a silver plate. And Rameses got the Hittite king's daughter as his wife. The peace treaty had eighteen points and was written in two languages and rigorously observed in their lifetime by both Rameses II and Rameses III.

Let's have another look at the graph (fig. 4). Three peaks of warming, which are equivalent to worsening of the climate in Asia Minor and Middle East, are accompanied by three state outbursts there.

The first peak: Akkadian Empire rises (soon to be collapsed into the trough of the next global cooling, which for Middle East means humidification, i.e., climate improvement).

The second peak: Shamshi-Adad establishes Assyrian kingdom, and soon after, the same is done by his Mesopotamian colleague Hammurabi who defeated Shamshi-Adad's successor. It must be said that it was a very bad time for that area in the climate sense—it was the era of the lowest water flow for the entire history in the Tigris-Euphrates river system. Also at that time, a powerful Hittite state rises on the territories of modern-day Turkey. And this latter-day state proves very soon to be a not-too-puny aggressor—in about 1600 BC, Hittites got to Babylon and captured it.

The third peak (about 1000 BC): King David unites Israel and Judea, and the state of Urartu rises in the southern part of the Armenian highland. It was a significant dry spell there at the time, which is corroborated by the extremely low water levels of the Dead Sea and the Lake Sevan.

The same three peaks of warming correspond to the catastrophes of "antiphased" Egypt, when three Egypt kingdoms collapsed (the New Kingdom, just as the first two, collapsed during a very propitious climate period near the third peak—in about 1000 BC).

Somewhat earlier, in the late thirteenth century BC (about 1200 BC), the powerful Hittite state perishes under the onslaught of the "sea peoples"—Phrygians and Mysians who came from the territory of modern-day Greece. The capital of Hittites, Hattusa, fell precisely when it was warm and humid in Asia Minor, or very propitious. It was much better than in Egypt where Egyptians, embittered by foul climate, under keen leadership of pharaohs Merneptah and Rameses IV were holding the country, albeit it hardly, against the inroad of famished northern barbarians sailing up from the Balkan Peninsula that was also suffering from "foul climate."

Egyptian papyri describe in detail how, as if by magic, numerous peoples of the sea appeared suddenly sailing up aboard their ships and drifting in like locust—and being very aggressive!

However, pharaohs were able to hold against hordes of sea nomads for not too long. One hundred years later, in the region of 1,000 reading, when the temperature drop was all of a sudden replaced by a brief "thaw" bringing the Nile residents some profusion and relaxation, the Egypt Kingdom fell.

OK, but what was meanwhile going on in China? Well, the same. Against the background of dropping temperature, the first centralized state appears in East Asia—China of Shan. It rises at the same time as the New Kingdom is crystallized in Egypt. And the Shan power falls practically at the same time as the New Kingdom collapses in the late eleventh century BC. The reason is the same—it became warmer.

The next collapse of the Chinese unified state (it was the West Chow dynasty to fall this time) happened in 771 BC. Please note: on our graph, it's the very peak of global cooling. And it's confusing: if it's cold, then we should see states pulling together, centralizing, and strengthening rather than collapsing ... Is it an unfortunate exception? No, it's not. Regional paleoclimatic analysis shows that against the background of general global cooling, China was experiencing then an abnormal brief warming that lasted about one and a half centuries. And it was at this time, during the "abnormal regional" period, that the country fell apart, splashing more than a hundred fragments. Supporting this warming, for instance, is the fact that in the suburbs of the capital, Chang'an (modern-day city of Xi'an), the heat-loving Chinese plum (*Prunus mume*) was grown, which was the daily food for Chinese—they made rice—and meat-sauce from it. Today, this plum doesn't grow in the suburbs of Xi'an, since it's too cold for it here even at the times of our global warming.

In the end of the cold third century BC, the next unitary Chinese state is formed—the country's first ever empire of Chin. The most climatically unfavorable (droughty) part of China became the empire's "crystallization core"—the unifiers came from there. It's interesting that the Chinese Empire was born nearly at the same time as the Roman Empire. By the end of third century BC, when Rome was forging its empire through fire of Punic Wars, the Chinese ruler of Chin dynasty, Shih Huang Ti, forged with fire and sword the then seven existing Chinese kingdoms into one—his. After that, he started building the Great Wall of China in order to protect the country from northern barbarians. A little later, Romans would start doing the same—building great walls.

The Chin Empire was replaced with the Han Empire that was spreading about the map "in the cold," capturing part of modern-day Mongolia, Korea, and coastal China ... This monstrous formation collapsed at the same time as the first terrifying and scary crumbling signs came from Rome that was still basking in luxury—in the early third century BC. However, Rome was able to stand its ground then and managed to hold out for some time, while the Han Empire broke up into three states.

Since we touched upon collapses of empires at warming peaks, let's look at the graph from the issue of peoples' migrations, that is, inroads of hordes of savage nomads. "Repetition is mother of erudition ..."

Between 3100 and 500 BC, fifteen big migrations of people happened. They were recorded reliably in history and were either accompanied or not with military confrontations. At the cooling peak (it's the second from left on the graph), the savage nomads Hittites—who were from either Transcaucasia or Southeast Europe and in any case from the north—invaded the blessed Anatolia and founded their own empire there. At the peak of the following warming and prosperity, Egypt is invaded from the north by Hyksos driven by drought. A little later, driven by the same drought, Aryans rout the Indo-Harappan civilization. In general, the lengthy cooling trend lasting from 1700 through 700 BC gave rise to whole hordes of migrating people! At the intermediate peak of cooling (about 1500-1400 BC), invasion of Dorians destroys the Minoan civilization. At the 1300-1200 reading, migrants from Melanesia get to Fiji and move further into West Polynesia in search of a better place. At the same time, the Jews perform their famous Exodus from Egypt. The Bible describes eloquently the most diverse natural catastrophes forcing the Jews to quit the Nile valleys that were blessed until then.

Of the fifteen great migrations of people, all fifteen were caused by local climate worsening: thirteen by global cooling, and two by global warming. The latter two were settling of Phoenicians in the Mediterranean in the eleventh to ninth centuries BC and settling of Etruscans in Italy in about 1000 BC. Why did the warmth drive them from their old haunts? Because Phoenicians lived, if someone forgot, on the territory of modern-day Palestine, Israel, and Syria . . . And Etruscans, according to one of the versions, are descendants from Asia Minor where they inhabited Lydia or the neighborhoods of the city of Side. But we remember that in Middle East and Asia Minor the term *warming* is synonymous to *drought*, or climate worsening. And it was the drought, which drove Phoenicians to colonize African coast of the Mediterranean Sea and pushed Etruscans onto the plains of Italy.

In Greece, life was relatively decent then and Greeks were OK. But in two hundred years, closer to the eighth century BC, it turned so cold that Greeks started showing concern and spreading about the world like cockroaches.

. . . During an era of worsening climate conditions, excessive population for whom there's not enough food is literally puked by the homeland. This happened with Greece, Phoenicia, Egypt, and it happened with Europe that repeatedly threw its sons into the crusades and that later used its excessive population to colonize nearly the entire planet. But the excessive population in Europe remained nevertheless! This excessive resource had to be burned already in the twentieth century AD in the furnace of two world wars—in the same purposeless way as the associated gas is burned in flare stacks at oil refineries . . .

The famous Greek colonization was caused by the cooling of the Axial Age that we'll discuss later. But it's interesting to note that the Phoenician spreading stopped just before the start of the Greek colonization, because at times of cooling the Middle East gets better (more humid). The climate swing moved—and some relaxed while others migrated . . . The last spike is the echo of Phoenician settling—Carthage founding on the territory of modern-day Tunisia. The great city was founded in 814 BC by descendants of Phoenician city Tyre.

The reasons for rapid but brief climate cooling (up to three to four years) are usually volcano eruptions. Could it be possible that it were the large eruptions to serve as the trigger stones for people's-migration avalanches that were long-ready to break loose? It's easy to agree with this supposition when looking thoroughly at the following table (table 1). The more so when considering that the power of volcano eruptions shown in the table is one order of magnitude greater than the Krakatoa explosion.

Table 1

Chronological Comparison
of Mass Migrations and Strong Volcano Eruptions
in Second to First Millennia BC

Year BC	Event	Date of powerful volcanic eruption
1700	Hyksos invasion of Egypt	1695±74
1550	Aryan invasion of North India	1623±73
1450	Dorian invasion of Crete; fall of Minoan culture	1454±69
1200	Exodus of Jews from Egypt; "sea people" invasion of Egypt; Hittite state collapse	1192±64
1100	Phoenicians start settling around Mediterranean	1084±62

750	Greek colonists start settling around Mediterranean	737±55
700	Scythians start moving from Central Asia to Europe	737±55
612	Midian, Scythian, Cimmerian invasion; Assyria collapse	585±52

CHAPTER 3

We Climb up Steep Ridge— Just with Axial Means

(The chapter title is a quote from a song by a popular Soviet poet V. Vysotsky—Translator's note.)

Karl Jaspers, the German philosopher, called the era between the 800 and 400 BC the Axial Age. It was indeed an amazing period of history! The era of unsurpassed flight of mind and human activities. The Axial Age left its signs in every corner of the globe (assuming, of course, that globe has corners).

. . . The chief technological breakthrough of the Axial Age was invention of iron. The Bronze Age gave place to the Iron Age. From the high centers of culture, iron spreads into the most remote corners of Central and West Europe, Africa, and China. Iron plowing tools helped China greatly. Prior to cooling, they harvested two to three crops a year, using the most primitive tools. And after the cooling, they needed iron for plowing. And not only that: it was in the Axial Age when the first draught cattle appeared in China. Construction of the Great Canal of China started, connecting the Hwang Ho and the Yangtze rivers. Implementation of a project like that could be only done by a unified country. Thus, fragmentation gives place to concentration. Ch'in Shih Huang Ti unified China into a single state.

By the way, iron also helped Negroes. It was only upon mastering iron when Negroes, forced to move by the changed climate, cut through the equatorial jungle belt into South Africa where black people were never seen until then and would probably be much scary. And who would not shudder?

. . . The main geographical breakthroughs of the Axial Age were voyages of Hanno and Egyptians around Africa. Both Egyptians and Phoenicians suddenly were itching to go on round-the-world voyages, and Egyptians used Phoenician ships and sailors to undertake a "round-the-world" voyage and in several years sailed around Africa clockwise. And a hundred years later, Carthaginian Hanno sailed around Africa counterclockwise.

Having passed through Gibraltar Strait, the Carthaginian navigator sailed along the west coast of Africa and got all the way to Cameroon. His feat impressed contemporaries so much that the detailed report on the voyage was engraved on a memorial obelisk in Carthage—in essence, the logbook. When Romans captured Carthage, the obelisk was destroyed along with the city, but fortunately, historians who had visited Carthage earlier wrote down the wonderful text enabling it to reach us with all its details.

While the "round-the-world" voyage of Egyptians had a rather scientific and cognitive nature, the scale of the event undertaken by Phoenicians clearly shows that it was not simply a voyage but a very real migration of a people in miniature. Sixty fifty-oared ships and thirty thousand people sailed off from Carthage in the late sixth century BC with the purpose of finding new homeland. And many found it—Hannoans founded several colonies on Africa's western coast.

Hanno describes how they passed through Gibraltar Strait and went along western coast of Libya (Africa), founding settlements one after another. Occasionally they went deeper inside the continent to survey the area. And everywhere on their way, they saw rivers and lakes overgrown with reed, hippopotamuses, elephants, crocodiles . . .

"Having set sail and passed the Pillars, we sailed for two days and then founded the first city that we named Phoemiatyre, close to a big plain. Then we sailed toward . . . Libyan cape covered with dense forest . . . and then went east for half a day until we reached a lake located near the sea and filled with lots of tall reed. There were elephants here and many other animals that were grazing. We sailed along the lake for one day and inhabited seaside towns called Karpkontikh, Gitta, Accra, Melitta, and Aramwy. Having set sail from here, we came to the big river Loukkos flowing from Libya. Local Lixus nomads tended grazing cattle there; we spent some time with them as friends."

—

Wherever Hannoans disembarked, they saw savanna, where the endless dead Sahara spreads today. But I wrote already that Sahara used to be savanna in the past, so let's not be repetitive.

The most important economic breakthrough of the Axial Age is the invention of money. Money in today's sense of the word. Humankind was generally moving toward universal measure of cost since long ago. And that path was not straight. Not many people know, for instance, that banks appeared long before money. Banks were in Mesopotamia in about third and second millennium BC—their role was played by temples and fortified palaces of local rulers. For a small interest, they provided merchants with services on safe storage of their goods.

In about 2250 BC, the rulers of Cappadocia took upon themselves the right and duty to guarantee the quality and weight of silver ingots, since silver served as the universal measure of cost. It was yet another step toward money.

Around 1200 BC, a new hieroglyph appears in China—"money," with kauri shells serving as money then. A little later, the same hieroglyph signified not shells anymore, but ingots of various metals in the form of knives, shovels, hoes, axes, which performed the function of money. An amusing evolution of money, isn't it? Metal has high value because tools can be made of it and well preserved and always exchanged for something else. And then gradually an inversion takes place—the exchangeability rather than functionality of metal tools comes to the fore. But for a long time yet, the money has the form of tools that are no longer used as tools.

And only in the Axial Age—in 640-630 BC, in Lydia (modern-day Turkey), money appears in the way we understand it today—round coins made of electron (gold and silver alloy) and standardized by the state. It gets easier as it goes on. In just another half century, real bank operations start. Some chap Typhus opens network of offices in different cities of Greece and Ionia while using noncash payments, because travelers don't carry money with them for security reasons and instead use promissory notes from the bank of Typhus.

During the same period of the Axial Age, China also starts to coin money of iron and introduces a term *nominal*, or "face value," meaning that the money was valued in accordance with not the amount of metal in it, but rather what was written on it. *Nominal* is in fact the "cost of power," or conversion of "trust in power" into real buying power.

First currency and financial experiments started. The Spartan legislator Lycurgus was the first in the world to introduce a policy of fiscal isolationism: in order to limit activities in his country of foreign merchants trading luxury items, he forbade the use of gold and silver coins. The money in Sparta was made of cheaper materials and therefore was huge. A more or less serious sum of money needed to buy large lots of goods was in no way transportable, and instead of carrying a small purse one would need a large cart pulled by an ass or a horse. Besides, the Spartan money was not convertible—it was accepted nowhere else besides Sparta.

Now let's change the topic from finances to culture. We see an upsurge here too! It's in the Axial Age that written language appears in Mexico—tremendous cultural breakthrough for civilization. In Greece, an Olympic movement is born—the first Olympic Games take place. Homer writes his *Iliad*. Hesiod and Sappho create next to him. In India, Upanishads are created. And generally, the entire hemisphere sees the golden age of creative work and science—plays, poems, and dramas are written, and amazing discoveries are made . . . The founder of natural philosophy, Thales of Miletus; Pythagoras; Socrates; Aristotle; Archimedes; and other well-known Greek philosophers and historians are all offspring of the Axial Age. Density of events and concentration of persons of outstanding spirit in the Axial Age are simply stunning. Nothing close happened earlier or later.

And finally, the Axial Age is the time for discovering and cognizing the individual human personality. It was at that time that practically all world religions were born, as well as foundations of contemporary morality. Siddhartha Gautama—the founder of Buddhism . . . Zarathushtra—the founder of Zoroastrianism. Mahavira—the founder of Jainism. Lao-tzu and Confucius . . . Moral sermons of Socrates . . . It's not for nothing psychologists and historians believe that the Axial Age was that period when the very type of personality was formed, that is the major type populating the planet today.

And it was not for nothing that Jaspers selected this particularly amazing period in the humankind history. The two eras, one to the "left" of the Axial Age and one to its "right," are not as densely saturated with scientific, cultural, and other significant events.

And the reader should not be surprised anymore that this remarkable time indicated by Jaspers (who for obvious reasons wasn't acquainted with methods of modern paleoclimatology and never saw our graph) falls on the cooling peak that was one of the strongest in history of humankind. There was only one cooling stronger, which happened more than two thousand years before the Axial Age—it served as the trigger to human civilization.

The Axial cooling left signs not only in material nature—glaciers, annual rings, isotope composition of ice and lake deposits—but also in human memory. The book of the prophet Jeremiah, written allegedly in about 600 BC, exclaims, "... Lord says: ask among the peoples whether anyone heard something similar to this? Does the Lebanese snow quit the mountain rock? And do the cold waters flowing from other places dry up?" That is, the unthinkable things are listed in this small quote—never before snow would quit the Lebanese mountains! And cold water from melted snow never stops flowing from mountains. In other words, generations of people at that time saw the Lebanese mountaintops always white. That is, they could ski even in summer. But they had no mountain ski resorts then, people were uncivilized and knew nothing of the beauties of expensive ski-passes and fast car-wings. Today, there are mountain ski resorts in Lebanese mountains. But they have problems with snow: it does fall in winter, but totally melts away by summer.

Here's another notable episode from the Bible. The book Proverbs of Solomon reads: "If chill from snow comes during harvest, it is the ambassador true to its sender for bringing joy to its master's soul." Can you imagine Palestine? And what about snow in Palestine during harvest? It was apparently so common then that it was used in proverbs ...

And here's what Theophrastus wrote about Crete climate in the fourth century BC: "As Cretans state, winters are now more severe and snowy—mountains were once inhabited and yielded grain and fruit to those who sowed and cultivated the land, but now vast valleys among mountains are deserted because they are barren. Once ... they were inhabited and the island was populous because rains were plentiful and snow never fell."

Theophrastus is seconded by Herodotus, who writes about climate in Crimea, the famous health resort and forge (of longevity) in the USSR, "... the winter is so severe that the unbearable cold lasts eight months. At this time, no matter how much water you pour on the ground, there would be no dirt, unless you build a fire. The sea and entire Bosporus of Cimmerians (that was the name of the Kerch Strait then) freeze up, enabling Scythians living on this side of the ditch to set off on a march over ice and use their carts to get to the other side, to the land of Sindhis. Such cold weather lasts in those countries eight months in a row, and it's not warm in the other four months."

Herodotus was often accused of being a storyteller and a liar! The eighteenth, nineteenth, and twentieth century historians did not believe Herodotus. But why refer to historians close to our time when even his colleagues living just a couple of hundred years later didn't believe him—Herodotus was harshly criticized by Strabo and Polybius who lived in a better climate time. And besides, it's impossible to believe in such descriptions of Crimea!

But Herodotus was right. And it's not for nothing he is called "father of history." This ancient historian did never invent anything but scrupulously wrote down everything he saw with his own eyes (he traveled a lot) and what others told him. This kind of ingenuousness did occasionally cause the old man trouble. But on the other hand, it allows proving today that he was right.

Sometimes Herodotus candidly records his doubts about what he was told, although he writes the story itself with no changes. He makes that kind of a comment, for instance, when describing the earlier-mentioned journey of Egyptians around Africa during the reign of Pharaoh Necho. In about 586 BC the Egyptians, who did not like to sail on the seas, hired Phoenicians for a specific task—to reach the Fairyland Punt and if possible to sail to the edge of the Earth. Phoenicians, being experienced seafarers, put all wad in a sack, got on board ships, and sailed off. The edge of the Earth—OK, let it be the edge of the Earth!

Three years passed. Egyptians thought by then that the expedition was lost, when all of sudden the boys came home. They sailed off in the Red Sea and returned in the Mediterranean to the mouth of the Nile. The journey took so long because the seafarers pulled in to the shore several times, got their ships out, repaired them, sowed grain and waited to harvest the crop, thus refilling their supplies. And went on with their journey after that. It's possible to grow two crops in eight to ten months in Africa, so they didn't have to wait long.

So Herodotus, who was told about the pioneers' feat, could not believe it because it contradicted the ancient Greeks' views on the world structure. The Greeks believed the equator could not be crossed, because they knew that the further south you go, the hotter it gets. They assumed by extrapolation that there were such countries where it would be impossible to live: the air is so hot that it's impossible to breathe it; the sea boils, making it impossible to sail on it; and the sun burns the ship sails. In other words, there exists the equatorial zone that is in principle insurmountable for humans.

"I myself don't believe in this journey, but you—suit yourselves," wrote well-educated Herodotus. And as an argument why he doesn't believe it, he noted that at one time in the journey the travelers saw the sun in the north, and that is complete nonsense! But it is this particular fact, which indeed proves the truthfulness of the journey, since the sun can be seen in the north only in the Southern Hemisphere. Such nonsense as sun in the north could not possibly be invented—it could only be seen personally! Phoenicians saw it and told Egyptians who then told it to Herodotus. Herodotus did not believe it, but honestly wrote it down. A marvelous old man!

Oh well, I'm digressing again. Let's get back to climate.

Dramatic events were happening in Europe then. Archeologists discovered signs of those events long ago, but they didn't quite know what to make of them before the appearance of paleoclimatology. Back in the mid-nineteenth century, pile-dwellings of ancient people were discovered in Austria, Germany, and Switzerland. Since Neolithic times, people settled along the shores of rivers, bogs, and lakes, making their dwellings on piles. So they were settling there for some time, but then in the mid-first millennium, they suddenly stopped. People deserted all their settlings because of a catastrophic flood—the water level of the Lake Bodensee, for instance, rose then at once by 9 meters (30 feet). The number of mountain settlements in the Alps has also decreased, even in those places that people would not leave willingly—next to metal and salt mines. But on the other hand, they could not afford not leaving: glaciers were advancing and blocking alpine passes. All this speaks to the fact that it not just became colder, but the precipitation rose sharply, too.

In England, plains were quickly turning into swamps. So quickly that during several hundreds years of the Axial Age the peat layer in one of the swamps in Wales gained a whole meter (3 feet) in thickness—the same amount it gained in the next two millennia. In order to save their roads, local citizens began planking them with logs, making in essence log-roads. But soon even log-roads would not help. And it became so bad that local citizens had to change their mode of transportation and get into boats.

The water level in lakes rose, but the World Ocean level dropped, which happens, as is known, in times of severe cooling—ancient sea harbors built in that era lie today more than one meter (3 feet) under water. When the Axial Age ended and climate started to warm up, the World Ocean level began rising. Ultimately, Romans even had to leave their old port at the mouth of the Tiber and build a new one in Ostia 10 kilometers (6 miles) upstream from the old one. Flood covered not only the old port, but also all the old Roman salterns located at the coastal edge. The same thing happened to salterns in Greece, causing for a time even some salt shortages. It was during the Axial Age that Greeks started wearing warm woolen clothes and flat roofs of the Minoan period were replaced by gable roofs (so dear to a Russian heart)—to prevent snow accumulations.

. . . If trying to keep up with the modern trends you decide to buy property in the Mediterranean area—Italy, Croatia, or Montenegro—keep in mind that Europeans are buying more houses today in the mountains rather than at sea. It's cooler. Most valuable are the houses in beech groves. For the sake of these groves, German couples of senior ages go up high in the mountains to altitudes of 1500-2000 meters (5,000-6,500 feet), where they breathe the forest air. But in the times of Rome's founding, there was no need to go up in the mountains to breathe the beech air, because Rome was surrounded with beech forests all around. Beech is a cold-loving tree, and when climate warms up, beech forests go up in the mountains where it's cooler. Roman peasants, being unaware of contemporary ecological whims, were only happy when this happened, for the beech trees were replaced by more useful grapes and olives.

Chinese chroniclers of the Axial Age record climate horrors similar to those in Europe. In China in the first millennium BC, summer frosts and snow were recorded for the first time ever in observation history (and in ancient China, sky observations were done thoroughly). Late in the third century BC, strong snowfalls and freeze killed many people in the capital of the Chin kingdom. The Yangtze River began freezing up in winter, and if someone forgot—it flows in subtropics. It should not be surprising, since the outlook of the Great Plain of China has drastically changed at that time. The areal of bamboo forests was pushed several hundred kilometers (miles) southward. Peasants were no longer harvesting two crops a year. In the remaining forests, heat-loving species of plants and animals have vanished completely.

On the other hand, it was the opposite in northwest India, where global cooling of the Axial Age led to drying up—precipitation at that time was 50 mm (2") lower than today. The army of Alexander the Great was leaving India after suffering great losses from deaths in sun-scorched deserts of Makran—the areas where many rivers flow today. Natural cataclysms are always shocking to people. And that's not surprising: it's hard to remain cool when looking in the face of death. One starts involuntarily thinking about meaning of life, immortality of soul, and other rubbish. Poets plunge into poetry, mystics into praying, and philosophers into jabber . . .

People perceived powerful cooling as the coming doomsday. It's not without reason that in Scandinavian sagas Ragnarok appeared—legend of the end of the world and death of gods, as well as a story of the endless winter, which narrates that "snow flies from all directions with freeze-burning wind; three such winters follow one another with no summer between them." Similar stories about endless people-killing winter were born at that time on the Iranian highland and foothills of the Altay.

The world has changed a lot then—economically, psychologically, and politically. In countries with a long history of thousand-plus years, like Egypt, priests tried to find in their centuries-old annals, something similar to the nightmare happening then, but they found nothing. And nothing could be possibly found!

. . . Unprecedented time. Unprecedented finds of mind . . .

But strictly speaking, times like the Axial Age happened more than once in history of humankind. Many an "axis" such as this one occurred. If taking a thorough look at the temperature graph in the area of 1800 BC, it's possible to see how sharply the temperature dropped. In just about a hundred years, the climate system tumbled down by almost half degree (almost 1°F). It was catastrophically fast and catastrophically much. It was painful. And for that reason, it was very fruitful because no success comes without suffering—while learning to walk a child falls and gets bumps. So let's look at humankind successes in pre-Axial cooling. The density of cultural and political events is once again quite impressive.

In China, the first high city-culture appears—Shan. In Greece, the famous Mycenaean culture is born. Aryans destroy the Indus culture.

Hyksos are driven out of Egypt. In Europe, the first linear writing appears—in Crete, proto-alphabet is born when pictures and pictograms (birds, insects, hands and feet) are replaced by conditional signs or symbols. It was a qualitative leap! The next step in abstraction would be alphabet. In India, the oldest Indian religion is born, forerunner of Hinduism—Brahmanism. A literary work, the first ever in the history of humankind, is written—the Vedas. The first monotheistic religion appears. I should dwell on this a little.

A religious reform is carried out in Egypt in the fourteenth century BC during the reign of Pharaoh Akhenaten: all old gods are directively abolished as not equal to the job, and the cult of a single god is introduced—Aton. Aton is the god of the visible solar disk. By the way, the pharaoh-reformer was previously called Amenhotep IV. The name Amenhotep has the meaning "Amon is pleased." Amon is the ringleader of the former gang of gods. The new name the pharaoh made up for himself and by which he remains in history—Akhenaten—means "pleasing to Aton." From now on, Aton is the new and the only god! And Amon gets a kick in his butt.

This amazing story of religion reform in Egypt is still a puzzle for contemporary historians. They still can't comprehend the essence of what happened. Well, it is strange indeed. There is a thousand-year-old religious tradition. A new pharaoh ascends the throne—Amenhotep IV, the son of the great Amenhotep III . . . He was great indeed! He reigned almost forty years and the period of his reign was the golden age of Ancient Egypt. Memories of Amenhotep III survived him by thousands of years. If you ever visit St. Pete, look at the faces of stone sphinxes on the Neva River shores. Their faces are that of Amenhotep III. Forty years of great pharaoh's reign are marked by huge construction—primarily temples and primarily honoring the chief god, Amon.

So now, the son of the great pharaoh ascends the throne. His pharaoh titular environment includes the old god Amon. And nothing initially would point to a religious storm. And what could possibly cause it? The atmosphere in the family of Amenhotep III, as well as in the higher echelons of the country, was religion-tolerant for those times. One of the teachers of the future religion reformer was originally from the Crete Island. And Cretans were by that time hardly even religious and had quite a light-minded attitude toward religion if not even atheistic. Europeans—what can you do with them!

It's a different story in the peasant Egypt. Peasant masses are very rigid. If present-day Middle-East Muslims are ready to kill for innocent cartoons printed in European media, illiterate Egyptian peasants were unlikely to have a more democratic mood toward subverters of their father's faith. People staged riots on lesser occasions, and here was a whole religion revolution! But Akhenaten-Amenhotep IV succeeded in it. In mere several years, he overturned 1,500-year-old views. Why did religion-tolerant Amenhotep IV become so unindifferent toward religion all of a sudden? So unindifferent that he risked insulting the utter Egyptian peasants over their deepest feelings. And most importantly, why did the reform succeed? And how was it carried out?

First, an official image of the new god appears, which is drastically different from the former images of gods: Aton, the god of the solar disk, is pictured as a circle with each of its many rays having a hand at its end. The difference from former images is its ultimate abstractness—no zoomorphism or anthropomorphism. No human bodies, bird heads, etc. Just a circle. And rays. And underneath, a small sinusoid underlines the circle.

The pharaoh assumes a new, sunny name. He founds a new capital whose name, Akhetaton, translates as Aton's Heaven. But that's not all! The reform rolls ahead with the inevitability of a road-roller.

In the ninth year of Akhenaten's reign, the name of the former chief god, Amon, is deleted from all signs.

In the twelfth year of his reign, the pharaoh generally forbids to use the words *gods* and *god* as directly associated with the former composition of the heavenly politburo. Only Aton! And both the sun and the pharaoh now have the same name—"sovereign."

And what about Egypt? Did it rebel? Did demonstrations roll along the Nile in defense of unjustly insulted faith of ancestors? Not at all. Both the people and the priests calmly swallow everything. Moreover, those whose birth names were given after the former gods are now hastily changing their names. They choose most often the old name of the Sun God—Ra.

. . . Indeed, all these oddities must have a very weighty reason!

In the seventeenth year of his reign, Akhenaten dies. And his reform is gradually curtailed. First, the former god, Amon is worshiped once again; then in Thebes, the word *god* is brought back into use. At the same time, the new Sun God, Aton is also worshiped.

Another quarter century passes, and Akhenaten is now cursed and declared a criminal, while his name and names of his "accomplices" are deleted and erased, just as it was done before with the name Amon.

The reasons for this full circle return are obvious: thousand-year-old traditions prevailed. But why was it so easy for Akhenaten to break the bundle of thousand-year-old traditions with a twelve-year-old twig? An external factor must have helped him. External and visual, like the war of the gods, in which Akhenaten was forced to take the side of the victor—the god of the visible solar disk. And nobody even argued with him—so obvious was the pharaoh's rightness. What could the sun fight against in front of the whole people? What could that be?

Egyptian (i.e., Pitch-) Darkness

A volcano on Santorin Island erupted then. Its location is not close to Egypt, about 120 kilometers (75 miles) north of Crete Island and 700 kilometers (435 miles) from the Nile delta. But the study of bottom sediments enabled volcanologists to restore the picture of grandiose catastrophe. The eruption we're talking about was a threefold serial. The core, which was extracted 130 kilometers (80 miles) southeast from the volcano, had the volcanic sediments up to 2 meters (6.5 feet) thick. Much less had reached Africa where the layer of ashes was only 1 mm (0.04") thick. Yet, this does not mean that the sun was not obscured over Africa. It actually was obscured because winds blew in the direction of Egypt.

It's interesting that the lower volcanic layer (the first series of the eruption) has red color because of the abundance of iron compounds ejected during eruption. This immediately brings to mind the Bible with its story about the waters in the Nile turning into blood—meaning that the water became red.

Egyptian Executions . . .

The Bible describes the war of the gods colorfully—water turning into blood, locust invasion, emergence of new diseases, pandemics, epidemics . . . This is all too similar to postvolcanic symptoms! A scientific work was published recently, which studies the largest eruptions already in the Common era, that is, in the last two thousand years. Based on historical documents, it shows that after practically every large eruption, unpleasant events of sanitary nature happened—pandemics of plague, smallpox, or other diseases, invasion of parasites ... And of course, crop failures. Because after every large eruption, temperature drops for several years.

If you're not too lazy, please look once again at the table of eruptions—1695±74, 1623±73, 1454±69, 1192±64, 1084±62, 737±55. When superimposing this chain of dates on the temperature graph, it's evident that each eruption methodically hammers toward cooling. And all of them lie on the lengthy—almost one thousand years long—cooling trend. This trend ended with the Axial Age. The 1695 eruption was the start. And just as the situation after that eruption began improving and the temperature was crawling upward, the 1623 eruption dropped the temperature like aspirin. And then another humble attempt of warming—and once again the 1454 aspirin (plus or minus half a century).

So we got to the reason for the Exodus of Jews from Egypt—crop failures after all those doomsdays and Egyptian executions. Until then, Jews had a decent life in Egypt—just as they had a decent life in Mesopotamia earlier. They enjoyed it. But then the situation worsens, and Jews customarily flee in search of a better place. The final push for their fleeing Egypt was the next eruption whose date, 1192±64, was determined after boring the Greenland ice. It added more cold, as well as resolve to brave Jewish lads to run following their nose.

. . . By the way, that was when famous Jewish monotheism was born—the cult of god Jahve. Jews are on the run headed by the chief runner Moses who teaches them on the way how to love the homeland. Or rather, not the homeland, since "homeland" for Jews is a variable value, but God! Their God replaces homeland ... Now here's a reasonable question: did Akhenaton's monotheistic religion reform affect the Jews? Did the Egypt's marginal people, the Jews, borrow their monotheism from aborigines? I believe it's something to think about.

But let's get back to Santorin. The volcano had three series of eruptions, which coincide with the three stages of establishing new religion by Akhenaten. Let's go through these stages in detail.

During his second year of reign, the pharaoh includes the following words in the list of his titles: "The only one for Ra," thus emphasizing special importance of the god of sun. The pharaoh's palace in Thebes gets an unusual name, "Castle of Exultation in Firmament." Perhaps, the pharaoh and his retinue exulted because the firmament had somewhat lightened up. Like, for instance, after the eruption's first series.

During his fourth year of reign, the pharaoh sharply intensifies campaign for "god of the visible solar disk," Aton. Firstly, the people are told that the pharaoh is the direct son of Aton, meaning that being his dad's relative, he can ask him unceremoniously to bring everything to order. The name "Aton" is now always preceded by a word "blessed" (or to be more exact, by the life symbol "ankh," which could be translated as "may exist forever," "may be glorified," "eternally alive," etc.). The earlier-mentioned official image of the new god appears—the sun with rays. And finally, the worship ceremony of Aton is fully formed, which is to be conducted not in a temple but right in the street—under the sunrays.

During his sixth year, the pharaoh assumes the new name, Akhenaten, and founds the new capital, Akhetaton (Sun City). And he declares from the podium (chariot):

"I shall create Akhetaton for Aton, my father, right here! Not south, not north, not west, not east of this place . . . but right over here! Because this was the wish of Aton himself, my dad."

This is a loose translation, as you understand, but the essence is clear: it's going to be in this spot and nowhere else, and as to why—don't even ask me, it's god's will! Just like that. Apparently, it was at this spot, when the darkness dissipated after the second eruption, through the black clouds, the son of Aton saw for the first time the crimson-red face of his father.

. . . How joyful it was!

During the ninth year, destruction begins of signs with the name of the former god, Amon. And during the twelfth year, elimination begins of the words *god* and *gods*.

In other words, events in heaven were developing as follows. Each of Akhenaten's steps in introducing the cult of the new god seemed to be confirmed by the heaven. Like Amon and Aton were fighting in heaven.

The sky got covered with pitch-dark clouds and the water in the Nile turned red. The sun was no more! "Was this not the result that we paid too little attention to the solar disk?" asked itself the political elite and first of all Akhenaten. "Should we perhaps pay more homage to the solar disk?"

No sooner said than done! And the sun appears! But the priests of the former Areopagus of gods are unhappy with the infringement of their position. And it's natural: not a single person fired is happy with being fired. So now, the pharaoh is helped by the eruption's next series. The sky is again pitch-dark. The pharaoh makes another push for worshiping the god of the visible solar disk. And the dark dissipates once again!

A while later, the old priests try again humbly to hold up their heads: "We're convinced, the new god, of course, is good, but we should not forget the old ones, your honor. Our fathers and grandfathers, after all . . ."—"I don't know, I don't know, well, maybe indeed we should not necessarily forget the old gods." And then it rumbles for the third time. It's cloudy and pitch-dark once again! And again the pharaoh rushes to his dad-sun in panic, assumes the new name (after his dad), and threatens to rip the butts of all old priests whom he has not yet had a chance to do so. And the dark dissipates again! Well, at this time, even the dumbest fool would understand . . . And the pharaoh understood. Enough! Enough fooling around! No more trying Aton's patience. The practice, known to be the criterion of truth, showed visually—not only to the priests, but also to the dullest Egyptian peasants—who was the master in heaven and who was his representative on earth.

". . . And I want to hear no more crap about any other 'gods'! I'm fed up with turning off the light . . ."

That's why there were no riots of dark religious masses and no old-believers burned themselves in some secluded monasteries. On the contrary, the Pharaoh Akhenaten enjoyed great popularity among his people. And the heaven itself confirmed it.

And only during times of Akhenaten's descendants, when a new generation grew up for whom the Egyptian (or pitch-) darkness was merely ancestors' myth, the opposition held up its head. The unfinished-off priests of old Amon and his gang were taking up the power little by little. Weren't they hit enough by volcanic bombs?

PART 4

At the Junction of Two Eras

I came to you to bring not peace but sword.

J. Christ

"It's somewhat too neat with these climate fluctuations, Vladimir Viktorovich," and I scratch my head. "Isn't the human history being overly deterministic by the temperature fluctuations?"

"And it would be strange if it were different. During thousands of years, up through the early twentieth century, the economic base of civilization was agriculture that directly depends on climate! However, occasionally the humankind helped the nature with own actions. For example, the army of Alexander the Great, numbering sixty-eight thousand troops went down the Indus on several hundred ships built by troopers from local wood. Apparently, troopers of the great conqueror finished off what was started by the cooling of the Axial Age—reduced to zero the relict forests remaining in those areas since the previous humid era . . .

"Much later, the natural catastrophe of desertification in North Africa was accelerated by Arabian nomads who came there in the tenth century AD. It must be noted that prior to their coming, North Africa represented a remarkable fragment of the Roman Empire, consisting of countries with up to 90 percent Christian population. As was customary, they fought one another, until one of the fighting nations got an idea—why not invite barbarians to crush the enemy? And they brought them upon themselves. Barbarians came and captured not only whom they were supposed to but all of the North Africa as well. And since these wild Islamites were cattlemen and nomads, they quickly cut down all the forests and orchards that were totally useless to them. This has accelerated the desertification catastrophe multifold.

"But responding to your confusion, I'll share with you a counter-confusion. I'm being totally vexed when reading works on history: these multi-tome publications have no mention of climate fluctuations!

But how is it possible to understand history and its motive forces with no knowledge whatsoever about climate changes in a historical era? Then what are you studying? Bare events? But what caused them? The same old crusades, for instance . . .

"Only when historians are pressed, they sometimes remember about climate—if it's totally impossible not to mention it. For instance, when writing about Central Asia cultures, they cannot escape it—researchers arriving in Asia hundred to hundred and fifty years ago and discovered dozens of big cities built in desert and buried in sands. Even the dullest historian would realize at this point that there was no desert here in the past—the cities were not built amid sands after all! Thus, the landscape changed.

"It should be said in all fairness that today, many scientists start thinking about the role of the natural factors in humankind history. I believe the first big incitement in this direction was made by Ellsworth Huntington's book *Civilization and Climate*, published in 1908. The book is quite useless, although it drew great interest enabling it to run into either six or seven editions in the next forty years. However, everything in this book was made badly! Huntington lived in that time when almost nothing was known about climate fluctuations. And to illustrate climate changes in Central Asia he studied annual rings of California sequoias! It was clear for some time that America and Central Asia is not the same thing. But today's level of knowledge tells us that the climate in these areas changes actually in antiphase! That's why everything in Huntington's book is wrong—simply rubbish. At least it's good that his books were not yet translated into Russian."

"And by the way, why were they not translated?"

"Because of Marxism that believes that everything dealing not with economy is from the Evil One."

"Useful thing—Marxism . . ."

CHAPTER 1

Icebergs in Bosporus

Greek lads were not the only ones who were observant—Romans were, too. They, too, saw the climate change! Such a notable agrarian as Columellia, the ancient Roman writer-bumpkin who left for his compatriots and progeny numerous volumes of works on agriculture, wrote: "Places where it was impossible before to grow any grapes or olives due to long and severe winter, now, with the warming and disappearance of former frosts, will be strewn with grapes and olives." And Columellia was referring to his compatriots' works written in the early first century BC. Columellia lived in the era of the emperor Augustus in the early first century AD. This means that in just one hundred years, the climate changed so much that viticulture became possible in Italy.

This brief warm era gave way to cooling, and by the end of the first century AD, the average hemisphere temperature dropped by almost 0.3°C (0.5°F). At the time of this brief cooling, Trajan was involved in a tough military campaign in Dacia, and to facilitate forced crossing of the river he built the famous stone bridge across the Danube near the Iron Gate. For this job, he invited the popular architect Apollodorus from Damascus. The bridge was built in the early second century, at the end of the cooling, and it stood 170 years until being swept away by the ice drift. In other words, during nearly two centuries, there were no serious freeze-ups of the Danube. Nor they happen today. But they were—just half century ago. This is indicated by data of 1950s-1970s that are traditionally considered the contemporary climate norm. And they are wrongly considered, because the Earth's climate warmed up significantly in the last half century—by as much as 0.4°C (0.7°F). And the Danube now almost never freezes up.

The graph below shows fluctuations of the average annual temperature for the Northern Hemisphere from 600 BC through AD 800 (fig. 5). Compared with the graph in fig. 4, this one is somewhat stretched horizontally and is more accurate. The temperature indicates the deviation from the climate norm of 1951-1980.

The highest peak of the graph is the time when the Apollodorus Bridge triumphantly stood across the Danube. Then it became colder, freeze-ups started, and the bridge was swept away.

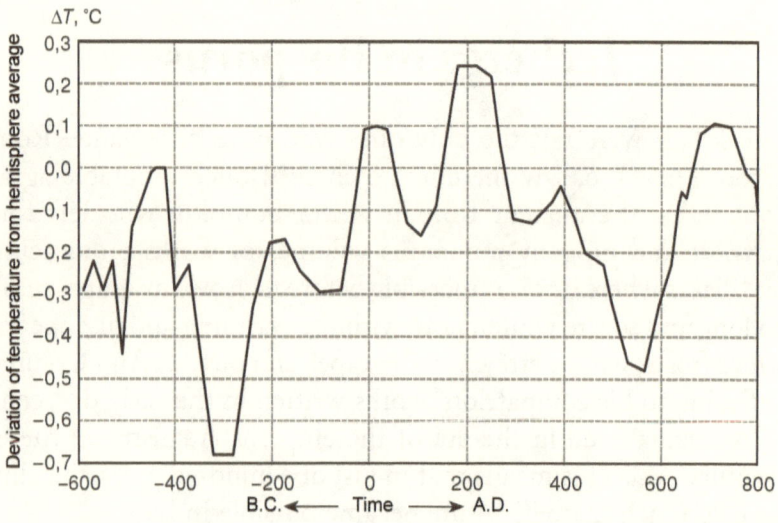

Fig. 5. Reconstruction of average temperature in the Northern Hemisphere from sixth century BC to eighth century AD

... In 1989, the Anatolia News Agency reported that the warming was causing the destruction of the glacier on Mount Erciyes (3,916 meters, or 12,848 feet above sea level), and the melted ice exposed the remnants of an ancient Roman temple of the early era of Christianity. The temple was carved out in the rocks close to the Turkish city of Kayseri that used to have a different name—Caesarea of Cappadocia. What does this mean? It means that the temple was built in a warm era when the rock was not covered with ice. The start of the era of Christianity was indeed an amazingly warm epoch. The world reached the temperatures of that time only today—in the era of global warming. And the rest of the time between these two global thaws, the temple was buried under the ice mass . . .

To the left of the highest peak we see a smaller one, and between them is a hundred-year-long cooling (approximately from AD 50 through AD 150). As we know, cooling causes droughts in Africa. And indeed, in about AD 120, the African provinces of Rome were struck by severe drought—not a single drop of rain fell there in five years. The consequences of that grand drought were observed personally by the emperor Adrian visiting African provinces in 128. To the right of the highest peak is the next dry period in North Africa. It started right after the temperature dropped precipitously from its record value (on this graph) and began its protracted fall from +0.2°C (+0.4°F) down to -0,5°C (-0,9°F).

Cyprian, the bishop of Carthage, wrote in mid-third century about the impending end of the world, and he used the ever-so-frequent droughts as one of his evidences. Victor Vita and Corippus mention the droughts in the fifth and sixth centuries. The authors of early Middle Ages—Johann of Ephesus, Procopius of Caesarea, and Zacharias of Mytilene—wrote about the unusually cold weather that set in the Middle East and Mediterranean in the first third of the sixth century. It turned so cold that it started snowing in Mesopotamia (!), and the summers were so cold that "the fruits did not ripen and the wine was more like grapes turning sour."

It was in this late-Ancient and early-Christian era, when the image was created in history and literature of the frozen-up Danube opening an easy raid route for the northern barbarians. Numerous heathens—Scythians, Slavs, Goths, Huns, to name a few—easily crossed the frozen-up Danube and carried out their forays all the way to the heart of the great empire, threatening Rome and Constantinople.

At that time it was not just the Danube froze up—other European rivers were frozen up for three to four months, too, like the Thames, the Rhine, the Rhone, the Maas . . . At times, the whole Black Sea froze up and icebergs floated for several weeks in spring through Bosporus, sometimes destroying Constantinople fortifications.

Climate troubles vexed the Chinese at that time, too. Fortunately, the science has tons of information on China. Contemporary Chinese scientists were able to find, summarize, and analyze several dozen thousands (!) of historical mentions of weather retrieved from official chronicles, travel notes, personal correspondence, or other surviving written sources. And today we know very much about climate in China at the start of the era.

We know that in one of the coldest decades, 280-290 AD, spring frosts in the lower Hwang Ho River happened ten years in a row. We also know that frequency of droughts, winter thunderstorms, and dust storms increased since 260 AD. It reached maximum in the sixth century, after which it began declining. This is a typical symptom of global cooling ... In this regard, it's not surprising anymore that it was in the mid-first millennium, when about twenty once-flourishing cities in Northwest China were deserted by their citizens. They were driven out by the drought accompanying the climate cooling.

During the era of "Six Dynasties" (420-589), glacier storages for food products were established in warm seasons at the ruler's court in the capital of the southern Sung Empire—Jiankang (modern-day Nanjing). They had no refrigerating units yet at that time, while regular supplies of ice from the north were also out of the question (southern Sung and northern Wei were fighting a war). Thus, the southerners had to have local ice. And this means that the climate at that time was much colder than it is today. In accordance with the climate norm of 1951-1980, the average winter temperature in Nanjing is +3.4°C (38.1°F), and it never fell below the freezing point in the entire history of instrumental measurements. Ice cannot form at that temperature. And it did form in the fifth-sixth centuries.

In that same sixth century, in northern China, the famous agricultural treatise appears, "Necessary Art for Ordinary People." The Chinese researcher Chu Kochen (a.k.a. Coching Chu), who was studying this work, noticed that the blooming periods of apricot, mulberry, and jujube were two weeks later than they are today. And this is another proof that the average temperatures then were significantly lower than they are today.

However, warming begins in the mid-sixth century, and the main problem of this era (Tang dynasty era) is not the droughts, but floods.

CHAPTER 2

Cold Perfects Mind and Strengthens Character

A tedious reader, believing neither in author's truthfulness or science objectivity, might ask:

"Well, you're saying that great events happen in eras of cooling, like spiritual achievements, migrations of peoples, and generally the rise in frequency of historic events. But history is a thing of continuity! In any moment of history, so many events could be enumerated that any theory would simply fall to the ground!"

It's logical. And that's why—to eliminate climatologist's subjectivism—this book uses expert evaluations of historians. That is, only those events are taken into account, which are considered significant by the historical society itself. And they were superimposed on the temperature curve.

So let's now temporarily stop the tricky questions and continue our journey through time.

And so, the Axial Age is over—the era of the strongest cooling in the last five thousand years. It was the era that gave people almost all world religions, basic philosophical and moral teachings, Greek-Roman culture, iron, money, and other goods and chattels of civilization.

The temperature initially rises irrepressibly. But then in about 350 BC it starts falling again, the brat. The cold period lasts approximately through 50 BC. And once again, cultural achievements start pouring like from cornucopia. This is the time, when the Alexandria library was created, the Pharos lighthouse was built—one of the greatest engineering structures of the ancient world, and the Mouseion—the prototype of the first academy of sciences—was born at the court of Ptolemy ... This cold period was the flourishing time for philosophy and culture in China. The first Chinese poet Chu Yuan, famous Confucians Hsun Tzu and Men Tzu, the great writer of ancient times Chuang Tzu—they all lived in the fourth-third centuries BC.

And since we've mentioned China, let's follow its history. The next brief cooling (50-150 AD) brought China lots of new stuff, too. We'll talk later about military-political perturbations and discuss now only technology breakthroughs. This was the time, when plow farming was widely spreading in China, the interchanging fields system appeared, and cultivation in beds was introduced.

Breakthrough also took place in metallurgy: a method was invented for actuating bellows with the use of a waterwheel. Immediately after that, the use of buckshee water force began spreading in agriculture, too—for grinding grain with the use of watermills. A little later, the water-lifting machine was invented, causing a small revolution in the area of land irrigation.

What's interesting here is that all the above-mentioned gadgets were invented during brief cooling era of the late first to early second centuries, but began spreading, noticeable only 100-150 years later, in the third to fourth centuries, according to historians. Let's look at the graph (fig. 5). We see that the third-fourth centuries represent the time of the next cooling. And between the two is the technologically stagnant heat peak when inventions did not spread. Who needs them? It became good even without them!

Even more interesting is that the cold era that gave birth to agricultural inventions and first mechanization led to gradual dissolution of slave-owning system. The era of climate worsening is the era of shortages: on the one hand, slaves are plentiful and they are cheap (because of empire expansion resulting from vast captures); and on the other hand, all those slaves want to eat, which is not quite convenient in view of failing crops. And what's good about a watermill? It never wants to eat!

The ancient world at the junction of ancient and new era is the sunset of the institute of slave ownership. In Rome, slaves are replaced by colons—freed-hired workers, and best Roman agrarian writers state with one voice that slave labor is ineffective. In China, a similar thing is happening: complaints about slave labor being unproductive appear in the treatise "Yan Te Lun" (81 BC). Roman and Chinese books write in unison that slaves working in state shops and fields bungle their job since they see no reason for working well. It's remarkable that in a more plentiful era the owners didn't give a damn about those reasons.

In short, in the end of the first century at the very peak of cooling, new forms of production appear in China that are more "capitalistic"—bondage lease spreads. Large landowners, or latifundists, start using now in their farms the labor of so-called pin-ko and pu-tsui. When literally translated from the Chinese, these words mean "guest" and "dependant." The etymology clearly points out that these are not slaves. Since having first appeared on the arena of history in the cooling era of the first century, these "guests" also became spread widely only in the third century—in the era of the next cooling. And in the first century when bondage laborers just started to appear, they even had different names—nu-ko and tung-ko. It's clear from the context of ancient Chinese sources that nu-ko and tung-ko are almost no different from slaves with respect to their status. But the new word and its corresponding concept were born! And in the next cooling era (third century), it got not just new sense but new sound jacket as well—pin-ko and pu-tsui. These terms mean not slaves but dependent landowners. In essence, they are serfs. They are not yet free, but they are no longer slaves. They have their land and thus an incentive to work. They cannot be sold, but can be handed down by right of succession or as a gift.

The attitude toward the unfree people is gradually changing—becoming more humane. In the second century, unprecedented legislative acts appear—Kuang-wu-ti, forbidding slave killing, as well as branding. Historians argue whether it was still slave-owning or already feudal period. And while they argue, we'll quietly note that edicts on slaves, adopted in the second century by Roman emperors Adrian and Antoninus Pius, were much like the Chinese. They punish masters for killing slaves and order cruel owners to forcibly sell their slaves.

The main cooling of our entire era, which happened in the fourth to sixth centuries, was not just beneficial to Chinese culture. The golden tree of culture blooms in India, too. It's the century of India's mathematics, astronomy, medicine, and literature. It's the time of creation of the Kama Sutra, Ramayana, and Mahabharata.

Buddhism rapidly penetrates into China and Korea, reaches Japan, and gets to Southeast Asia.

Foundation of Christianity strengthens in the Mediterranean and Europe. Saint Augustine writes his book *City of God*. The famous temple of Saint Sofia is built in Constantinople. The monks in Saint Benoit Monastery devise a notable ethic formula—"pray and work—that later became an ideological foundation of European economy and life.

Irish monks of St. Jonah monastery begin their much-hard mission of baptizing Anglo-Saxons and Picts, and Irish missionaries get into wild continental Europe where they establish a whole bunch of monasteries in Austrasia, Burgundy, Neustria, and Langobard Kingdom. And by the end of the cooling—at the junction of the sixth and seventh centuries—the world's last religion is born at last in Arabia: Islam. The temperature curve then was going up (for instance, return of Mohammed to Mecca was already in 630), meaning life was clearly looking up. But here's a question: Why then the world religion, born as the temperature began to rise, developed rapidly after the death of Mohammed when the weather was warm? Isn't this a contradiction to theory? No, if we recall that in places where Mohammed lived the climate worsening is in antiphase with respect to the rest of the world. With the warming, Arabia began to dry up and life sharply deteriorated, causing a fire-like spread of the new paradigm.

And while we're in the Middle East, let's see how this region was affected by earlier climate worsening resulting from global warming. Let's for instance look at the warming around the year 0 (fifty years left and fifty years right from that date). At this climate-worsening peak in the Middle East, a prophet is born in the city of Nazareth, and the birth of a new world religion is associated with him. At the same warming peak, his followers, Peter and Paul, develop their energetic activities.

Next rise in warming at about the year 200 gives birth to a prophet Mani in West Asia, the founder of the fourth world religion—Manichaeism. This religion was quite popular in the Middle Ages and reached its apogee in eighth and ninth centuries, becoming the state religion of Uighur Khaganate, and then it petered out.

What follows from this? Well, it's that all world religions without even a single exception—Manichaeism, Christianity, Islam, Jainism, Brahmanism, Hinduism, Buddhism, Zoroastrianism, Confucianism—were born and spread in the eras of climate worsening associated with climate either cooling or drying up.

Let's now talk about inroads and migration of peoples. Sharp and brief cooling in about 400 BC triggered the great Gaelic avalanche. Gaels suddenly took off from their long-inhabited lands (the Rhone and the Rhine valleys) and started their disgraceful doings by invading Italy and capturing Rome.

A few words about this sudden peak temperature drop (see curve 2 in fig. 4—around 400 BC). Historical chronicles describe that in the early fourth century BC, a sequence of severe winters befell the Apennines. A whole generation of Romans grew up without ever seeing ice in their life, and suddenly the Tiber was freezing up every winter—and this happened three winters in a row! Most likely, this anomalous peak temperature drop against the background of general rise was associated with a large volcano eruption.

Rapid temperature drop in the early fourth century BC could not be caused by just one reason—for it to happen, undoubtedly a whole host of several important climate factors had to have an adverse confluence. And that's exactly what happened: around 400 BC, volcano Okmok exploded in distant Alaska that was unknown to anyone then. According to the eruption power scale, which was discussed in chapter 3, this eruption had a full six and it would be enough to drop the hemisphere temperature by 0.3-0.4°C (0.5-0.7°F). But that's not all. At about the same time, solar activity began decreasing sharply. And this, as we know, leads to cooling, too. And finally, the water of North Atlantic all of a sudden became less salty. Without getting into details of the complex issues of interconnection between salinity of North Atlantic water and global temperature, please take my word for it that desalination of water in this area always leads to serious cooling. By the way, it's the same horror item, which was used in recent years to scare trusting average people and on which basis the crafty Americans crafted the much-talked-of blockbuster *The Day After Tomorrow*, showing New York all frozen up.

Also then, in the fourth century BC, the cooling briefly stopped and the Gaels calmed down for a while. But right after 350 BC, the temperature was once again rising and the Gaels' motion syndrome became acute due to the crop up of the population surplus (i.e., food shortage). Here's the result. The freed (of food) Gaels rout Macedonia, sack Apollo sanctuary in Delphi (279 BC), and cross to Asia Minor where they find sudden quieting, for as we remember, in times of cooling Asia Minor experiences climate improvement. Gaels settle in their new blessed homeland and give it their own name—Galatia. The famous Turkish football (i.e., soccer) team Galatasaray that recently won the

UEFA Cup is named after the district in Istanbul known for its longtime inhabitants—the Galatians, that is the Gaels.

The cooling of the second century BC forced the Sakas and Yechi to leave their domicile and move.

The temperature drop of the fourth to fifth centuries AD forced the starving Huns and Ephthalites to move.

The continued temperature drop in the sixth century forced the Avars and Slavs to move.

Please draw your attention to the graph (fig. 5). In the mid-third century AD, a prolonged cooling starts and lasts about four centuries. It's the era of the so-called Great Migration of Peoples. Everyone was on the move! Not just the small nomadic people who are traditionally on the go at the slightest drying up of the steppes, but also the barbarians from the woods inhabiting North and Central Europe.

The Great Roman Empire, already quite weakened by earlier warming (the golden age of the Antonines), was suddenly pressed hard by barbarians from all sides, mildly at first but then getting ever stronger. And the colder it got, the stronger was the onslaught of barbarians from everywhere. Germans were pressing from the north, Slavs and Suebi from the northeast, Alans from the east, Scots and Picts from the northwest, and Berbers from the south . . . The pressure was mounting, until the structure cracked.

. . . The Great Empire fell. Europe plunged into darkness for centuries to come . . .

Since we started talking about Rome, it would be useful to mention that the entire history of triumphant rise of this remarkable state happened in the cold era. Rome was founded in the cold. This city-state conquered all of Italy in the cold. Also in the cold, it won the First Punic War and captured Sicily. Then the Second Punic War was won and Roman control was established over Iberia. At the same time, Romans crushed Macedonia. Following it was the final solution of Carthage issue and Pompey's conquering of Syria. And at the very end of the cooling era, Caesar conquered Gallia.

Then comes the warm era, and Rome is shaken by civil wars. Republic collapses and monarchy is in fact established.

In the times of the next cooling (around 100 AD), the Roman Empire reaches its greatest expansion during the glorious reign of Trajan.

Then the hundred-year warming arrives with all kinds of liberalisms that always come with thaws, like granting Roman citizenship to everyone or devastating epidemic of smallpox.

Then there's the next twist of cooling. Diocletian tightens the screws rigidly in order to repair and rebuild the empire that's gotten loose and half-collapsed.

The further drop of temperature makes life so unbearable for everyone that the whole world cracks and collapses, following the Roman Empire that fell under onslaught of savages who came running from everywhere like cockroaches . . .

In the Part 3 of the book, I promised to discuss in some detail the military-climate swing Rome-Byzantium vs. Parthia-Persia. Watch my hands—a trick is about to happen . . .

It's catastrophically dry in Parthia, while in Rome it's warm and nice. Rome suffers a shocking defeat. Marcus Crassus, the victor over Spartacus, is routed in the battle at Carrhae. Parthia captures Syria, and then Parthians besiege Jerusalem and defeat the army of Marc Antony. The time is 53-36 BC.

Then the climate trend changes and cooling comes during the reign of Trajan. Crops fail in Rome. And though it's a bit cooler in Parthia, too, but it isn't critical for agriculture in Parthia because it has some beneficial rains. That is, it became nice. And Parthia loses on all fronts—Trajan captures Armenia, takes Babylon, Seleucia, and the capital of Parthia Ctesiphon, and reaches the Persian Gulf. Mesopotamia and Armenia are declared Roman provinces . . . Only plague, which killed half of his army, and the Jewish uprising in the rear, force Trajan to return Ctesiphon and Seleucia, which he could no longer hold. But soon afterward, Marcus Aurelius retakes those two cities. The time is 115-164 AD.

Then again the change of the scenery: 170-260 AD. It's warm in Rome, and that means good, while it's dry in the East. And so Mesopotamia and Armenia are back under the power of the East. The army of the Rome emperor Gordian III is defeated in the battle at Euphrates. The Romans are also beaten at Barbalissos and Edessa. And finally, in 259, the Roman emperor Valerian is taken prisoner by Parthians. Total flop!

AD 260-350—it gets cold in Rome and humid in arid Persia. Bad for Rome and good for the East. Roman emperor Marcus Aurelius Carus invades Mesopotamia. Co-ruler of Diocletian, tetrarch-emperor Caius

Galerius enters Armenia with his legions and routs Persians. Armenia and a large piece of Mesopotamia are once again in the hands of Rome. Armenia enters into alliance with Rome against Persia.

AD 350-400. A small temperature jump. Good for Rome and bad for arid zones. But the temperature jump is small and uncertain. This causes the military scale to vacillate, and it aims to move not toward Rome. The captured Mesopotamia is returned to Persia, and Rome is forced to cancel the anti-Persia alliance with Armenia. Persia and Rome divide Armenia—$^4/_5$ of its territory goes to Persia. In the very early fifth century, Rome and Persia conclude a peace treaty on joint protection of Derbent and Darial passes—the only ways leading north across the Caucasus.

AD 420-650. Temperature continues to drop. Now it's bad already for both Rome-Constantinople and Persia-Iran. The arid East cannot be saved anymore by higher precipitation. It's very cold! Crops fall. It starts snowing in Mesopotamia—who ever heard of such a thing? The system starts crashing—the war is going with variable success, using up the last resources, aiming at destruction. The "eternal peace" treaty of 532 AD is broken. Persia captures Syria and Antiochia. Soon after that, Constantinople retakes Syria back, along with West Georgia. In 561, a new peace treaty is concluded, fixing the former borders between the two empires. As it turns out, they fought for nothing . . .

And the temperature continues to drop yet. It's bad for everyone, very bad, and it's necessary, obligatory to fight, in order to burn down the surplus population. So now, Persia-Iran recaptures Antiochia, Jerusalem, Alexandria, and raids Egypt and Anatolia in a devil-may-care manner. The Persian shah Parvez Khosrau II takes eastern and southern provinces from Byzantine Empire. His armies advance to Constantinople. But soon the Byzantine armies, led by the great commander Emperor Heraclius, launch a counter-offensive and invade Iran, then capture the shah's residence and take back Jerusalem. In 628, the two sides conclude a third peace treaty. And once again, the terms are the same.

. . . The world is cracking under tension . . .

Let's now analyze the reality with respect to strengthening of state centralization and stimulating the imperialization in the eras of cooling.

Each global cooling was accompanied by rise or strengthening of at least one empire. We spoke already about Rome. It rose and began its victorious spread about the Italian Boot in the much-discussed Axial Age.

The Axial Age was generally rich with empires that grew like bubbles all around the Northern Hemisphere. For instance, the Eastern Empire of Achaemenids rose exactly at that time, too.

During the second cooling (300-20 BC), Rome spreads from the Atlantic to Syria.

In India at that time (around the end of the fourth century BC, the very start of cooling), Mauryan dynasty is born, which founded the first empire in the country's history.

In mid-third century BC in droughty Trans-Caspian steppes and semideserts, the Parthian Kingdom was born, lasting almost five hundred years. In about 200 BC, the Greek-Bactrian king Demetrius captures northwest India, and a bit later king Menander snips off another big chunk of India and thus spreads his influence over a lump of land from central Asia to Ganges.

In China at that time, after almost three-hundred-year-long internecine war (the Era of Fighting Kingdoms), the country is unified into Chin Empire that was replaced right off by Han Empire. Chinese emperor Wu Ti expands his country's borders west to the Tarim River and south to the Hainan Island, modernizes the Great Wall of China, and even establishes communication with the Roman Empire via Parthia.

The third cooling (AD 50-150). Rome is at the peak of its might. In Central Asia, the Kushan Empire strengthens and spreads from the Aral Sea to India. In East Africa, the Aksum Empire expands; although it's not known to many, it nevertheless not only flourished for over six hundred years, controlling all of the sea trade with India, but also swayed the destinies of peoples in the Red Sea basin.

The fourth cooling, which was almost as strong as the cooling of the Axial Age, is marked by a no-win deadly confrontation between two superpowers of ancient times—Byzantium and Sassanid Iran, which we have discussed already. But here's what the attention should be drawn to: Sassanid Iran was born in early third century AD from expanded Parthia, during the times of warming. But we remember that in those places warming is synonymous to climate worsening. Before being screwed together into a single country, Parthia was perceived by contemporaries (by Romans, for example) as a patchwork quilt, a confederation of several kingdoms (eighteen, according to Pliny). And each of these kingdoms had its own king at the helm. But in the era of the next climate worsening, the Sassanids turned the patchwork quilt into a monolith.

—

Historians know that the start of the Common Era was marked by general system crisis encompassing such remote and diverse states as Rome spreading from the Atlantic to Syria, huge China of Han dynasty, and Parthia spreading from the Amu-Darya to the Euphrates and the Red Sea. All these giants seemed as if they were struck down by some common social disease. The cause of this disease was perhaps the high-frequency climate fluctuations in the end of the past era and start of our era, when temperature was jumping up and down with amplitude of 0.2-0.4°C (0.4-0.7°F). Duration of favorable and unfavorable periods was about one hundred years. And the rise and fall of temperature happened in twenty-five to thirty years, or within one generation. In other words, climate fluctuations coincided with the "socium natural frequency." Not surprisingly, the system came into resonance and went into destruction.

In favorable periods, survivability of newborns surely increased and they managed to grow up to the age when they were able to bear arms, but the system was unable to feed them. Instead, it was able to throw them into the fire of war, thus eliminating in bad times the surplus human masses accumulated in good times.

The West (Rome-Byzantium) and the East (Parthia-Persia-Sassanid Iran), being ultimately played out by one another, concluded a peace treaty, third in succession, and breathed heavily like two animals with their tongues hanging out. And right then, a third one sneaked up imperceptibly, the one who always is the winner when two giants are fighting. Arabs! In the battles against them at the Yarmouk River and at Al-Qadisiyyah in 636 AD, the Sassanid Empire collapsed. And Byzantium, having lost the former Roman grandeur in a flash, turned into an ordinary common kingdom, becoming one of the many. And this formation of a new Arab empire happened, as you have perhaps guessed, in the era of temperature discomfort.

. . . As always. Time to get accustomed . . .

Let's now set our magnifying glass over China, for the reader was promised to discuss its military-political perturbations from the viewpoint of climate. So we're moving back in time . . . fifth to fourth centuries BC, the era of fast improvement of climate situation. The world was quickly rising from the cold Axial era to a warm one. And right then, unrest started in China, with centrifugal trends prevailing. Historians know this era as the era of Zhan-guo. The country broke up into seven pieces fighting with one another in languid military campaigns, until it began to cool again—in mid-fourth century BC the temperature curve changed its direction and started moving downward. And the colder it turned, the clearer it became that the military-political balance was tilting to the advantage of the northwestern border-state of Chin. In 316 BC, Chin subdued the first two kingdoms, Shu and Pa.

From bad to worse: the Han dynasty replaces the Chin dynasty and now rules over an incredibly huge China—from Manchuria to Guangdong. This monster ceases to exist in the next warm era—in 9 BC the power is seized by the regent of the juvenile heir to the throne Wang Mang. After that, a new era of unrests and riots sets and lasts exactly through the end of the warm time—late first century BC and early first century AD. And then everything ends as usual—a new strong chap comes and screws everything back together.

The chap's name was Kuang Wu-ti, and the empire he restored was named by historians as Junior Han. Temperature was sliding down, and China was spreading wide, showing in all sorts of ways its great-imperial ambitions. And it was spreading so vastly that nobody thought it was too little. The troops of the young empire hit hard on the northern barbarians' domes. The army of the Chinese commander Ma Yuan captured North Vietnam. Chinese general Tou Ku carried out a series of brilliant punitive expeditions, clearing the Great Silk Way from the Xiongnu, who became insolent during the internecine wars, thus restoring the empire's influence in the west. Two hundred thousand Xiongnu were brought into the empire as workforce. But most distinguished was the general Pan Chao whose troops reached Central Asia and got to the Caspian Sea.

. . . You agree—heart beats in joy when reading such stuff. Although we're not China, but it's still nice . . .

Pan Chao even managed to kick the butt of the earlier-mentioned Kushan Empire army, which expanded with the first cooling of the new era. Being the governor-general of the captured western territories, Pan Chao was also very active as a diplomat and sent a representative delegation to establish contacts in the west. And the delegation got all the way to the Mediterranean Sea. The Chinese ambassador arrived in Antiochia where he was received with all the honors.

However, the Junior Han Empire's triumphant march on the history arena does not last long—during the next warming it finds itself in the same place where it was in the previous warming periods: in the mire of civil and internecine wars. First, China breaks up into three independent

—

states and holds out for some time in this "suspended" mode only thanks to the new cooling that should awake centripetal trends. But then a brief heat burst comes in about 400 AD, and China, balancing on the edge, cracks and splits into twelve independent states.

When do you think the period of unrests and splintering ends? Let's look at the graph (fig. 5). You're right, in the coldest trough—in the very mid-sixth century, China again becomes a unified state, and ascending the throne is the dynasty with the name Sui. We postulated in the previous part of the book that China is an ideal model for studying the effects of climate on social centralization. And this was said not for nothing. The point is that China represents a "clean experiment"—due to its geographical location, very few external disturbances affect it, such as hordes of savage nomads who could have substantially amended the history.

Earlier we repeatedly observed some things seemingly contradicting our model—in cooling times some empires instead of strengthening were collapsing under external assaults of nomadic savages who simply had nowhere to retreat. Force beat force. As to China, it was threatened only by northern barbarians. And if we were to neglect this disturbance, just as mathematicians neglect infinitesimal quantities, the Klimenko climate theory would work in China with no failures: cooling—state concentration, warming—disintegration. Periods of unrest and civil wars in overwhelming majority of cases correspond to warming times. And even if some unrest does happen in a cooling period, such as for instance reformatory socialist movement of Gracchi brothers in Rome, it suffers defeat: you can't fool around with a strong state.

PART 5

Middle Ages

. . . In the heated fights and woes,
Humans boil with tempers storming,
As each hour comes and goes,
World is eaten up by Warming.

While this eating never stops,
To the final gulp it races—
When the ocean of "hot spots"
Swallows islets of "warm places" . . .

Boris Vlakhko

"Anyhow, my friend, it seems to me that it would be hard for many to believe in the strict determinacy of climate over history. Their arguments are simple: 'This is how climate, even most severe, affects me—I put on my fur coat and hell with it.'"

"Climate affects not just the outward appearance but also national traditions and character. Konstantin Pobedonostsev, Russian jurisprudent, politician, and teacher of the future emperor Alexander III, once said, 'Do you know what Russia is? It's an icy desert where dashing men roam.' Do you know where the true border lies between Europe and Asia? It's the same line as the winter freezing-temperature isotherm—eastern Poland, western Ukraine, northern Black Sea coast . . . To the left of this isotherm are Europe, West, Catholicism, and Protestantism; and to the right are Asia, East, Orthodoxy, and communality. This borderline is easily visible with a naked eye: in winter, to the left the land is black and it rains, and to the right it's all snow-white.

"Why is Europe afraid of Russia up to now? Because Europeans—Poles, Balts, Germans, and others like Hungarians believe up to now that Russians are barbarians from snowy woods. But Hungarians, for example, are closer related to barbarians—they are direct descendants of savage nomads who came to Europe from the depths of Asia a thousand years ago. By the way, there's a historic province in Hungary called Cumania, Cumans being the ancient name for Polovtsians (a.k.a. Kipchaks). They were common Asians from Trans-Ural and Pre-Ural steppes and clear representatives of Finno-Ugric tribes—narrow-eyed, short, and bandy-legged. Their closest relatives in present-day Trans-Urals are Khanty (Ostyak) and Mansi (Voguls).

—

"Ancestors of present-day Hungarians invaded Europe in late ninth to early tenth centuries during one of the cooling periods that chased away nomadic tribes from their historical homeland. They were ordinary nomads, like Huns or Avars . . . By the way, the most popular male name in today's Hungary is Attila, and the central street in the Hungarian capital Budapest bears the same name of the person who inspired terror in fifth-century Europe. But did you see Hungarians? Do they look like the narrow-eyed and short Khanty and Mansi? Superficially, nothing of the kind! Even the narrow eyes disappeared somewhere. But they speak practically the same language! So that's what climate did to them. For sixty years, Hungarians inspired terror in Europe all over, ravaging it like plague. Hungarians captured Rome, went down to Italy's south and France's north, until in 962 German King Otto I beat them in the battle at the Lech River in the territory of present-day Bavaria. Otto foresightedly executed every single chief of furious Magyars, and Hungarians settled after that in plains that are called today Hungarian Plains. In just one hundred years, Hungarians became civilized, dropped their nomadic habits, adopted Catholicism, and created a centralized state. They transformed from nomads into Europeans. And they consider us Russians, living to the right of the freezing-temperature isotherm, eastern barbarians.

"A similar transformation occurred with Normans. They, too, inspired terror with their inroads all over Europe for more than a century. Normans had flat-bottomed ships enabling them to sail over both sea and rivers. They sailed deep into Europe and captured Ruan, Paris, Rome . . . They were absolute savages! Normans committed totally senseless and useless atrocities—they crushed and destroyed everything they could not carry away and killed everyone they could not take along with them.

"These marine nomads sailed all the way to Greenland and America. And in the ninth century, Normans' indomitable passion for robberies brought them to North Africa where they used their inherent resolve and cruelty to kick out civilized Arabs from many strong points. In the eleventh century, the Pope of Rome hired these uncouth bumpkins to settle his own scores. By that time, Christian Church was already divided into Eastern and Western Churches. What happened in 1054 was the notable mutual damnation of the competing organizations: Catholic Church and Orthodox Byzantine Church. The pope decided to use savage heathens to kick out the Orthodoxes from the last fragment of Byzantine possessions—Sicily.

"Normans did kick out the Byzantines, but also gave the pope such a hard time that he felt he got more than he bargained for—in just several years the Normans dictated the pope their will and established their own kingdom in the captured Sicily (and had no intention of giving it back to the pope after capturing it). Just several decades sufficed for the dirty, funky, savage Normans to get civilized and build a powerful kingdom lasting several centuries and producing world-renowned rulers, of whom the most eminent was undoubtedly Friedrich II Hohenstaufen, the King of Sicily and at the same time the Holy Roman Emperor. His sarcophagus is in a cathedral in Palermo and buried in flowers every day up to now.

"Entire Europe called him Stupor Mundi—Wonder of the World—that's how highly erudite and wise was this fellow. He towered above the medieval Europe like a great pyramid of Egypt over desert. He was fluent in several languages, but spoke German imperfectly. He corresponded many years with Genghis Khan and Mameluke sultan of Egypt Al-Kamil, the latter correspondence being in Arabic. The two rulers discussed news in astronomy and geography. The Court of Friedrich was full of most eminent scientists of that time. By the way, this was the time of crusades with Christians fiercely killing Muslims . . . But Friedrich's authority was such that he managed to get an agreement giving Christians Jerusalem without war or redemption, with the Lord's coffin and the path to it from the sea. The agreement was signed, but unfortunately, it did not survive those who signed it. But the wars between Christians and Islamites stopped for twenty years, and free access of Christian pilgrims to the Lord's coffin was restored.

"In fact, this descendant of savage Normans was recognized as the leader of the entire Christian world. And at the same time, he was not a religious fanatic while his religious tolerance reached such a level that once being in Jerusalem he visited the main Muslim mosque of Omar (Dome of the Rock). It was an unprecedented step for that time. For the first time in history, the leader of the Christian world paid a visit to an 'enemy' temple! But Friedrich didn't care two hoots about those prejudices—his main task was to establish peace.

"He was a great man, whose greatness was recognized by everyone in the world of that time. When in 1250, Friedrich II died, nobody wanted to believe in his death. A legend even was born in Europe, alleging that in reality Friedrich did not die but just retreated with his troops to one of the alpine caves in South Germany, from where he would reappear if

Germany were threatened. This legend stayed alive up until our time, and one of its adherents was Hitler—his residence Berghof in the Bavarian Alps was built on the very same mountain where according to the legend Friedrich's cave was.

"As you see, one or two centuries of living in favorable climate suffice to civilize yesterday's dirty savages dressed in beasts' skins and to bring them to quite a decent condition. At times, this is reflected most favorably even in appearance: the Hungarians' Mongoloid features have evaporated as if never been present.

"If I remember correctly, Gumilyov wrote that pure-blooded Polovtsians are tall, blue-eyed, fair-haired . . . The old man wrote lots of funny stuff. I don't know how he got such a strange idea into his head, but it gets not just into *his* head. I visited many countries and remote corners of the world and heard almost everywhere from local small and dark aboriginal population that 'true' aboriginals are tall, fair-haired, and blue-eyed. They even say about Genghis Khan that he was tall, red-haired, and blue-eyed . . .

"It varies sometimes—'red-haired'! I too hear these stories constantly—in Turkey, Abkhazia, or Siberia . . . Even Jews sometimes say the same about themselves! Everyone wants to be tall, fair-haired, and blue-eyed. And no one wants to be a bandy-legged Mongoloid or a hook-nosed dark-haired Semite."

"But not everyone was as lucky with climate as the Scandinavians were. They are the only people whose ancestors were indeed tall with light brown hair and blue eyes. As to nomads that kept raiding Europe—they were common nomads being of Central Asia by origin. And by the way, if you read history books, you wouldn't note that when speaking about the raids of the nomadic people, the authors always mention that they were by origin from the foothills of the Altay Mountains. Avars, Huns, Tocharians, Hephthalites, and Yechi—all these successful conquerors were by origin from the foothills of the Altay Mountains. And the Altay is far from being the most favorable place on Earth: hot summers and very cold winters . . . Even today very few people live there. But it's in this place where people feel the severest climatic stress. The Altay is one of the most climatically sensitive areas on Earth. Any global climate fluctuation, even not very significant, causes here an impressive response. That's why the Altay generates conquerors in such monstrous numbers."

CHAPTER 1

The Gray Wolf Is Just Beside
and May Bite . . .

(The chapter title is a quote from a popular Russian lullaby—Translator's note.)

The second volume of the Academy issue of "History of Europe" enlightens its readers: "The climate of Middle-Ages Europe remained cool and quite dry up to the fifth century, but then, and especially in the sixth and seventh centuries, became more humid in dry areas and dryer in humid areas. The All-European warming lasted from eighth to thirteenth centuries, causing southern flora and fauna to spread north. Marshes were slowly disappearing in the south, being replaced by numerous meadows. The warming was moving southeast from northwest, reaching its maximum in Greenland by the tenth century, in Iceland by the twelfth, in Netherlands by the thirteenth, and in Russia by the fourteenth. As a result, the thirteenth century became "Golden" for West-European farming (in some counties of England vineyards were even laid down then). But then came the sharp cooling that steadily intensified up to the seventeenth century . . . The sharpest social consequences of nature's mess occurred between tenth and eleventh centuries, when Europe was shaken by a series of deadly earthquakes and other natural calamities coinciding with the anxious anticipation of the end of the world in the year 1000, causing unusual panic in many places."

This quote has practically everything wrong. Modern paleoclimatology shows an entirely different picture that could be observed by the reader in a graph below (fig. 6). The temperature shown is its deviation from the 1951-1980 mean value. The curve was plotted using all paleoclimatic data available today, including palynology (analysis of fossil pollen composition), limnology (analysis of lake deposit composition), dendrochronology (analysis of tree ring thickness and wood density), glaciology (analysis of fossil ice composition and movement of glaciers), history, isotopes, and phenology.

What strikes the eye at the first look of the graph?

First, feverishness of the up-and-down rushing curve. Second, narrow, or short in duration, warming peaks and the opposing similarly narrow but more numerous cooling peaks. Third, starting in 800-900, a clear general trend is seen toward cooling.

And what can be seen on the graph at the second, more careful look?

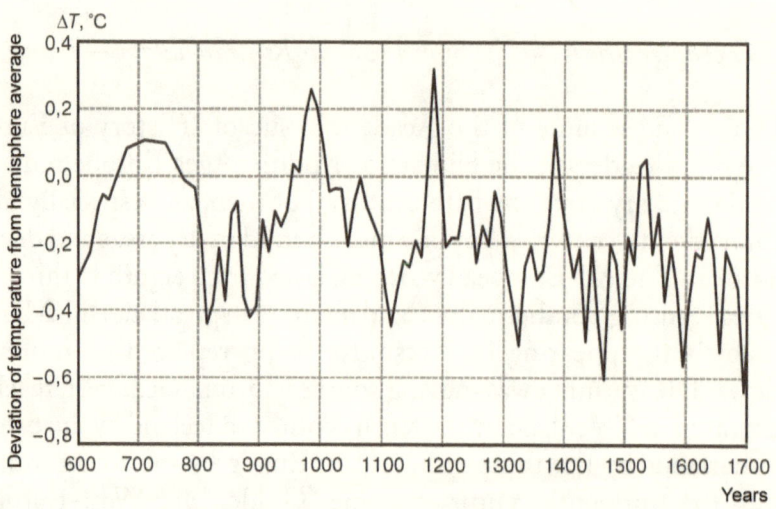

Fig. 6. Reconstruction of annual average temperature in the Northern Hemisphere during seventh to seventeenth centuries AD

The line that seems at first sight as a zero line has in fact a reading, -0.2°C (-0.4°F). In other words, except for rare narrow warming peaks, all in all the Middle Ages climate was noticeably colder than the climate norm of mid-twentieth century (1951-1980).

We know that cold times give rise to wars of conquest and to strengthening of states, while warm times cause disorders and unrest. That is, there are two sorts of military activities—centripetal and centrifugal. This theory is confirmed splendidly by practice: two thirds of the Middle Ages had severe climate. The remaining one third was warm and accounted for 65 percent of all centrifugal events contributing to decentralization of power—riots, popular unrests, revolts, civil wars, internecions . . . Think—just one third of the time produced two thirds of destructiveness!

Generally, it's strange—why in climatically good times with prospering life and high crops, people begin to riot, revolt, and overthrow power?

Psychologists and sociopsychologists have noted long ago that revolutionary crises happen not when people are poor and hungry but on the contrary, when they are replete, wealthy, and . . . unhappy. The paradox is that various revolutionary situations and crises are usually preceded by periods of economic growth rather than downfall. Unrests in society start not when "working people experience poverty and distress above the normal level," nor when the economic situation is bad with respect to objective indices, but on the contrary—when the economy is growing! Because people's expectations are growing at the same time. And since people's needs and expectations grow faster than the economy, a discrepancy occurs that raises the sense of dissatisfaction, when people start believing that their life is not what it should be and that their existence becomes simply unbearable. What happens is called in psychology a retrospective aberration, or semantic overturn: despite the objective indices showing increased living standards, people believe that everything is terrible and that last year life was better. And that's the way this situation is described by memoirists and chroniclers, giving out their own feelings for the actual situation . . .

It becomes especially dangerous if economic growth for some reasons gives way to recession or simply slows down, because expectations continue to go up by inertia! And a revolution is close by. A revolutionary explosion becomes very often associated with an unsuccessful war that was supposed to be victorious and small. And when a small war suddenly becomes not only not victorious, but also not small, social frustration develops with search for culprits.

For example, just before the French Revolution the living standards of French peasants and artisans were the highest in Europe. The same situation was in Russia and Germany in the early twentieth century—these were the most dynamically developing countries of the world with economic growth rate exceeding 10 percent a year. And at the same time, they constantly made rows!

In the very first years of the twenty-first century, a work was published by a historian who analyzed racehorses' names in prerevolutionary Russia. Terrorist, Bomb, Barricade—those were the names given to horses then! The general sense injected into the society was "May a heavy storm hit stronger!"

Now let's take for example a period of warm and favorable climate in late fourteenth century—Europe was then simply shaken by civil unrests! France is shaken by Jacquerie, Tuchins, and Mayonets. North Italy is shaken by revolts of Tuchini and Ciompi. In England, Wat Tyler commits excesses. The cities in Portugal, Germany, and Flanders are beaten up by civil disturbances, too. Ideas of disintegration are boiling in the minds of people. However, it's worth noting that not all rebellious actions were unambiguously bad and destructive. For instance, the period of iconoclasm in Byzantine Empire, which happened in the warm era, or the schism of the Christian Church, which took place in the warm era, too—were they negative phenomena or not? Divisive, decentralizing, disintegrating, rebellious—yes. But were they negative?

Now let's move along the graph (fig. 6). First, let's observe the temperature rise that reaches maximum in the first half of the eighth century. It's warm! Since dark Middle Ages reigned at that time in Europe, history did not save for us abundant written evidences of that era when the kings were illiterate and shamelessly picked their nose while wiping their green finger on their pants. And in China, it was the reign of Tang dynasty—people were already cultured, which is why there are numerous Chinese documents about that warming. In North China, there were occasions when two winters in a row were totally snowless and the Emperor's orchard in Sian (and it's far from subtropics) was full of not just plums but citrus fruit, too, like tangerines and oranges. Today, neither citrus plants nor plums grow in that area, as was mentioned earlier.

However, Chinese emperors enjoyed tangerines not for long. In about the year 800, the average global temperature dropped sharply, and the cold stayed about one and a half centuries. The start of temperature drop was in the cold winter of 763/764. Reminiscences of that nightmarish winter were recorded in not just Chinese chronicles but also in Byzantine and European surviving chronicles. In France, all winter crops were destroyed then by frost, which also killed olive and fig plantations on the Adriatic and Aegean coasts.

It was not just the European rivers froze up; the Black Sea up to 100 miles from the northern coast was covered with ice up to 10 meters (33 feet) thick. In spring, when all this ice began melting and breaking up, the winds pushed it down south, totally plugging up the Bosporus and even damaging the fortress walls of Constantinople. The future notable historian Theophanes the Confessor, who was eleven then, saw how people used the solid ice bridge to cross from Europe into Asia.

Later, similar winters would repeat with a depressing frequency. For example, it's known as a fact that in the mid-ninth century, the Adriatic Sea froze up more than once. In winter 860, the commodities were brought to the Venetian port over sea, but for the first time not by ships but by sleighs. This mode of delivery soon became customary and no longer surprised the population. The last severe winter of this kind happened in 873/874, after which the global warming started. The two-hundred-year-long, warm era began, during which the temperature exceeded the climate norm of the mid-twentieth century by 0.3°C (0.5°F). However, the temperature during this thaw was not always that high and had occasional troughs (after 1050).

I want to draw your attention to the heat peak at about 1190 and the subsequent temperature drop. The rate of temperature drop was huge—in quarter century it dropped by half degree (1°F). And climatologists consider a drop of 0.2°C (0.4°F) in ten years an ecological catastrophe because the ecosystems can hardly bear it—they simply don't have enough time to adequately react to the rapid changes. Let's keep this in mind.

The first one-third of the thirteenth century was simply catastrophic from the climate viewpoint. Frosts were killing vegetation all around Eurasia. And not just vegetation suffered, but great plans as well. In the 1217/1218 winter, an unusually strong snowfall ruined the Baghdad offensive by the armies of the Central Asia ruler Khorezmshah Muhammad. Being just 300 kilometers (190 miles) from Baghdad, the armies of Khorezmshah encountered a big trouble: so much snow fell suddenly that further movement of military units became simply impossible. Baghdad was saved and the downfall of caliphs was put off by another forty years.

However, the picture soon changed. When looking at the graph, it's possible to see that after the abysmal cooling at the start of the century, the climate of the entire thirteenth century was quite stable (the graph shows minor temperature fluctuations) and not too cold (about 0.1°C [0.2°F] colder than the climate norm of mid-twentieth century). Except for the horrible first one-third of the century, Europe practically had no cold winters. In fact, it was quite the opposite! Albert the Great in his "On plants" treatise lists grapes, olives, pomegranates, and figs among plants that were grown in the lower Rhine valley. And these were not some selected hothouse exercises of some eccentric feudal—no, these were plants that "grow in abundance in Cologne." Figs don't grow in Germany even now, although the climate today is warmer than it was then.

What does this tell us? Of the warm winters, dear readers, of the warm winters. They were so warm that in winter of 1289/1290 in South Alsace, trees did not even drop their leaves, strawberries ripened in December, and grapevines bloomed in January. In Vienna in the same year, violets bloomed in the end of December and apple trees bloomed in mid-January.

Similar warm winters happened that century in China, too. Administration of the new ruling dynasty Yuan even restored a special department charged with the oversight of bamboo plantations that along with citrus plants expanded their natural habitat vastly northward. However, this department did not exist long: in the end of the thirteenth century, temperature once again dropped sharply and the need in oversight of bamboo plantations was gone, just as the plantations themselves.

But it was not everywhere as nice as in Europe or China. For instance, in Russia, Poland, Czechia, and Siberia, summer seasons became significantly cooler and rainier. In Switzerland, Scandinavia, Caucasus, Urals, Altay, Himalayas, Tien Shan, Cordilleras, and New Zealand, the glaciers became active, which also tells us that the summer temperatures were becoming lower.

As we already mentioned, the fourteenth century began with a strong cooling. Europeans were by that time so used to warm winters that in England they were discussing in earnest the possibility of expanding farming areas for vineyards. However, the fourteenth century put an end not just to these talks, but in general to viticulture in Great Britain. Ever since then, no one was hearing anymore about English wine.

Subsequently, Europe experienced such frosts that many were thinking of those times as the start of the end of the world. The winters 1303/04, 1305/06, 1307/08, 1310/11, 1313/14, 1316/17, 1318/19, 1321/22, 1322/23, 1325/26, and 1327/28 were so severe that the Baltic Sea totally froze up, the Adriatic Sea began to freeze up again, and the Mosel and Rhine rivers were frozen up for four months.

In the first half of the fourteenth century in Germany, April and May snowfalls and frosts became frequent and destroyed crops, causing famine. But May is nothing! On June 30, 1318, snow fell in Cologne! Summer floods, becoming more frequent, were washing away crops from the fields, which also did not help prosperity.

Similar junk was noted in China, too. Diaries were discovered of some traveler named Kuo Tiang-hsi from the province of Chiang-Su, who wrote that he was forced to stop his sailing along the Grand Canal, connecting the Yangtze with the Hwang Ho, because the Canal froze up. Even the Taihu Lake, located close to Shanghai, was frozen up, which happened only twice in China history—in 1111 and 1329. And all this mess happened in subtropics! And there were more Chinese nightmares. In the very midsummer 1349, on July 31, snow fell in the then Chinese capital Kaiping. The later generations learned about it from the Chinese poet of the fourteenth century Nai Hsin. He wrote that swallows come to Beijing in late April and leave the city in early August. That is, their stay on the latitude of Beijing was almost three weeks shorter than it is today.

In the first one-third of the century, the Amu-Darya in Khorezm would freeze up for 4-5 months a year. For comparison, in the early twentieth century it froze up in its upper course for only 2 months.

Atlantic ice attacked the coasts of Iceland. In 1306, 1320, and 1321, the island was blocked by ice even in summer. In 1321, Iceland was hit by such a strong blizzard accompanied with freeze that almost entire livestock was killed, causing famine. Chronicles tell us the story: "The year of great need in Iceland; people everywhere are starving to death." It was since those times that the Icelanders renounced being involved in agriculture and instead turned to fishing.

If things like that happened in Europe and China, it's easy to imagine the situation in Russia! "Frost is strong in the entire land of Russia"—this refrain was often repeated in chronicles of that time when describing the cold weather in summer.

But the most horrible years for Europe were 1313-1317, when summer incessant rains and cold weather led to such severe crop failure that cannibalism started in Europe. That's when ogre characters appeared in West European fairy tales . . .

Misfortunes never come singly. After the famine, the weakened population was then struck by epidemics. In Russia, entire villages died out from Antonov fire—poisoning from ergot that affects rye (it happens usually in very wet weather). And in 1346-1353, Europe was hit by bubonic plague that reduced European population by one third. The plague came from Crimea—it was brought to Italy by retreating Genoese who were earlier defending the city of Kaffa (Theodosia) that was besieged by Janibek Khan of the Golden Horde. And the plague got to Crimea from China, having apparently traveled along the Great Silk Way. The point is that in 1332, catastrophic floods occurred in China, causing mass migration of rats—the carriers of plague. That year, floods and plague killed seven million people in China.

However, in the second half of this accursed century, global average temperature began moving upward, and global cooling was thus replaced by global warming, albeit a brief one. It lasted about thirty years, but wasn't unnoticed. Historical chronicles of England, France, Germany, Netherlands, Poland, Czechia, Greece, and Russia noted the mild winters and very dry hot summers. The Sun spots could be seen by a naked eye, which the *Nikonov Chronicle* honestly recorded: "The Sun was like blood with black spots on it, pitch-dark, . . . heat and sultriness were extreme, woods and marshes and earth were burning, and rivers were drying up, and other water areas were totally dry, and great fear came, and all people were horrified, and it was great grief."

. . . Both the frost is bad and the heat is not good . . .

In the course of thirty years, from 1361 to 1390, Russia had seventeen droughts. The same trouble hit Europe, too. After heavy spring floods, such heat set in that not just small rivers but many big ones nearly totally dried up.

After this brief warming, which brought no one any good, came a new cooling—even stronger than the previous one. The fifteenth century was an awfully cold one with just a few warm years or seasons.

During this century, snowfalls were observed in tropical areas of China (1415, 1449, and 1488). Today, the average January temperature in these areas never drops below +13.5°C (56°F). Rain with snow fell even in the southern part of the tropical Hainan Island where average January temperature today is +19°C (66°F). If things like that happened in the tropics, what is there to say about the subtropics! During the fifteenth century, the freezing up happened numerous times on the Yangtze tributaries and the Lake Taihu. At the turn of the fifteenth to sixteenth centuries, snowfalls in the Yangtze delta lasted five months, creating a snow cover in subtropics up to 1.5 meters (5 feet) thick, and all tangerine trees were killed by frost.

The same happened in Japan. Not far from Tokyo is the sacred Lake Suva with a monastery close by, hosting Buddhist monks. The monks clearly had nothing to do, so they conducted annual weather observations, motivated by religious reasons to do this job that was so useful for later generations but totally useless for them. Those monks in the fifteenth century noted very early dates (compared with today) of the lake freezing up. Just once in all of the fifteenth century the lake didn't freeze up—in 1457/58 winter.

Rui Gonzalez de Clavijo, the ambassador of the Castile court in Iran, wrote in his travel diaries in 1404 about huge amounts of snow that he discovered in this southern country. The severe winters did not spare anyone and spoiled the plans even of such a titan as Tamerlane. By that time, the lame old man conquered all of Central Asia and Asia Minor, Egypt and Transcaucasia, defeated the Golden Horde, marched to India, and now was preparing to fulfill the dream of his life—conquering China. But in November 1404, snow began falling, and it was so heavy that all preparations for the war and all construction work in Samarkand had to be stopped. The snow kept falling and falling with no intent of stopping. The roads that Tamerlane planned to use for the China offensive were buried under snow. The intelligence, sent by the Great Lame to the Karatau passes, reported that the snow cover there was two lances thick. The huge military machine of Tamerlane stopped. The waiting lasted many weeks, until finally Tamerlane got sick and died.

No less adventurous weather was in Europe. German literature names the fifteenth century as the era of desolation (Wustungen)—whole villages were dying out, the earth was lying lonely and empty. In West, Central, and East Europe, according to various estimates, from 20 to 60 percent of villages died out then.

The Adriatic Sea began to freeze up again. The witnesses wrote, "The snow was falling all month long. All the vineyards perished on the continent. People were arriving in Venice . . . on horseback over thick ice. When looking from Venice in any direction, snow could be seen everywhere . . . Ships were moved from one shore to the other by ropes."

Obviously, if the Adriatic Sea froze over, the Baltic Sea simply turned into smooth and firm ground that is very convenient for military actions.

The Swedish historian Olaus Magnus, who lived in that era, wrote a treatise "History of Nordic People," and one of its chapters was dedicated to the war on the Baltic ice, depicted in the following quote. "Battles on the ice of the Gulf of Finland in winter happen as often as in summer, when the Muscovites violate the terms of peace. The ice is so strong that it can hold the masses of cavalry and infantry . . . Horses have special shoes with nails to prevent them from sliding on ice."

In 1495, Muscovites under Ivan III attempted to conquer Vyborg by storming it from the ice of the Gulf. It was in November, which means that even in autumn the ice was so thick that it could hold the whole attacking army! And by the way, in the twentieth century even in the coldest winters the firm ice formed in the Gulf of Finland not before mid-December, as was the case during Soviet-Finnish war. But that year, when the angry Muscovites attacked the Swedish fortress Vyborg, the peasants of Pomerania traveled to Danish islands Falster and Moen by sleighs, easily crossing over 200 kilometers (125 miles) of sea expanse. And the German historian Albert Krantz, who lived in Lubeck then and wrote a whole tome with a distinctive title *Vandalia* on history of Slavs, noted: "The sea freezes up so firmly that it is possible to move across ice to Denmark and Prussia, and in some places there are even . . . inns established on ice for the convenience of travelers."

Packs of wolves ran over the North Sea ice from Denmark to Norway, searching for game and terrorizing the population with their foolhardy raids on villages and suburbs of European cities.

. . . It's since that time the European fairy tales have the Wolf eating the Little Red Riding Hood, and Russian folklore has the Wolf biting little kids at their side when they go to sleep. What an archetype . . .

Europe was successfully destroying its surplus population in senseless wars and was thus feeding wolf packs—no one who stayed alive would risk collecting bodies of those killed in action, because at night the fields filled with corpses were also full of green lights of wolves' eyes. Even the Duke of Burgundy Charles the Bold, killed in the battle of Nancy on January 5, 1477, was eaten by gray wolves. His companions-in-arms were not able to save the body of their commander, because the night fell—the time of wolves.

If it was freezing-cold and wolves howled in Germany, it's easy to imagine what was happening in East and even Central Europe! A German student Butzbach who happened to be in Bohemia (Czechia) described this country as "the kingdom of eternal cold."

—

And a total nightmare was happening in the coldest country in the world—Russia. Of the hundred years in the fifteenth century, forty were years of failed crops, or years of famine, and fifteen years of those forty had no crop at all. And it's not surprising. This is what the chronicles wrote.

In 1420, snow fell in Russia in mid-August, it was falling non-stop for three days, and it accumulated up to one meter high (forty inches)! The entire crop was literally buried.

Next year the frost started in September. On September 15, snow fell again, and it was falling incessantly for two days. Then, fortunately, snow melted and was gone, and people rushed to harvest the beaten-down ears of grain, but they didn't harvest much because in a few days snow began falling again, accompanied by freeze. Famine hit Russia in the next three years. All the dogs, cats, and crows were eaten. People were eating their own children. The city of Novgorod has died out by half. People were leaving their huts and going to nowhere. Majority of them died on their way from starvation and freeze.

In 1435, frost hit in June and killed the crop. Next year, frost hit in early fall during the harvest and killed nearly the entire crop again. A couple of years later, in 1438, incessant rains rotted the standing crops, and once again came famine and cannibalism.

In 1439 on May 1, snow fell up to knee-high.

In 1442 in spring, snow fell again and frost hit, and then a severe drought happened in summer. And once again, there was nothing to eat . . .

As the chronicle described it, "One could only hear weeping and sobbing in streets and markets, and many fell and died of starvation, children in front of their parents and fathers and mothers in front of their children, and many left—some to Lithuania, others to Latvia. At the same time, there is no fairness and justice in Novgorod where robberies started in villages and counties and cities . . ." Naturally, famine causes orgies of gangsterism. The chroniclers note that in those years gangsters were hunting exclusively for goods in kind and "took only bread, butter, millet, and corned beef."

Everyone suffered in the fifteenth century, in both Russia and Europe. And in Europe, by the way, it was the height of the witch-hunt season . . .

If there are too many people in a certain area, or in other words, if there's not enough food for all, the surplus population must go somewhere, self-liquidate. In times like these, people eat one another; kill one another in Holy Wars and liberating campaigns; change on a mass scale the area of their habitat for a one with more food; and start various "purges" like hunt for witches or heretics.

The peak of witch-hunt in Europe happened precisely in the fifteenth century. But witches existed before that. It's known, for instance, that the first witch was burned in Toulouse back in 1275. But at that time, it was a single event, because the church authorities, being the better-educated part of the population, didn't countenance such accusations. And they couldn't, the classics were against it: one of the pillars of church, Saint Augustine, never recognized any witchcraft and other superstitions like copulation of a human with the devil.

However, in the fifteenth century life became so tough for humans that they started destroying witches on a mass scale with direct blessing from and active participation of the church. It's worth noting that one of typical accusations was "weather spoiling." In his bull on December 5, 1484, the Roman pope himself recognized the weather spoiling as fact and launched a new wave of auto-da-fe. After that, fires were burning all over Europe for another half century.

And the wave of burnings was stopped only by the next warming, but not by criticism from humanists like Erasmus of Rotterdam, Andrea Alciati, and others. The Era of Reformation came along with the warming of the early sixteenth century. It was very brief, but relatively strong. As soon as the weather improved, the witches stopped being burned, and that practice was renewed only after 1560 with the start of the next wave of cold weather. And the witch-hunt never stopped since, until the end of the eighteenth century. Because the so-called Small Ice Age started . . .

CHAPTER 2

Crystallization of Intellect and Empires

We'll be doing now what we've already been doing earlier—follow the effects of eras of global cooling on flights of human spirit, intellect, and other achievements.

Huge Mind Can't Be Hidden Under Hat . . .

Let's begin with the very start of the graph (fig. 6). The first global warming of the Middle Ages is in seventh-eighth centuries. Everyone is fine, while arid areas are in big trouble. Accordingly, the arid zones were the place where the most powerful and cultured state existed in that time—Arab Caliphate. Its intellectual achievements are well known to all: Arabic numerals, astronomy, lunar calendar, medicine, algebra, and trigonometry . . . Arabs gave a lot to the world. It was in this cold era that construction was done of the famous Dome of the Rock (Mosque of Omar) in Jerusalem, the Grand Mosque in Mecca, and the Umayyad Mosques in Damascus and Haleb.

At the same time, Europe's cultural life was at the very zero. Sheer illiteracy with even kings and dukes unable to write or read, unused to wash their hands, and pots with their urine splashed out into fireplace causing terrible stench around the palace . . . The only cultural oasis in Europe at that time was Islamic Spain.

The first renascence of Europe begins in the mid-eighth century and lasts through the end of the ninth century. Historians call that period "Carolingian Renaissance." At that time, the entire territory of

West Europe belonged to the empire of Charles the Great, who cared for various sciences and arts and assembled around him on that basis a group of some fine-minded and well-educated chaps headed by some Alcuin of York. The group was called the Academia. The era of Charles and Carolingians gave birth to a whole constellation of scientists and thinkers: poet Angilbert, scientist Sedulius Scottus, philosophers Scotus Eriugena and Rabanus Maurus . . .

The king's court paid special attention to ancient heritage and to organizing new schools. Writing local chronicles developed immensely. Political treatises were created and literature was born. It was the era when Romanic and Germanic languages were formed. New easy-to-read writing was invented—the so-called Carolingian minuscule. Special shops at monasteries were involved on a mass scale in rewriting ancient books, because Charles's court started collecting a huge library. It was during that time the foundations were laid of European feudal arts. And at the same time, the construction boom began, too. They were building, of course, not houses for the poor, but cathedrals, monasteries, castles, and even swimming pools. The impression was that as if someone has suddenly built a fire in the darkness of the Middle Ages.

The reader, used to the paradigm, could probably say already, without even looking at the graph, that this was related to cooling. Indeed, global cooling in the end of the eighth century was the reason for the cultural flight of the era of Charles the Great.

It would be strange if the same didn't happen in England. The cultured King of Wessex appears on the island, Alfred the Great, who not only creates his mighty bunch of thinkers, but also is keen on writing—the king personally translated into Old English the "Consolation of Philosophy" by Roman poet and politician Boethius.

Similar cultural processes are boiling in Central Europe, too: in Moravia, Cyril and Methodius create the Slavic written language; Bulgaria and Serbia adopt Christianity. The language brings Byzantine missionaries to Kiev. (*Author paraphrases a Russian phrase, "The language will bring you to Kiev," which means that talking and asking could get you even to the capital city, Kiev, called mother of all Russian cities—Translator's note.*) In 867 in the mother of all Russian cities, the first Christian church is built.

In China, this cold period was called the golden age of Chinese poetry. In this time, creativity was high by such poets as Du Fu, Po Chui, and

Chinese Pushkin—Li Po nicknamed by his contemporaries "celestial being who was driven away to earth." Schools and libraries grow like mushrooms all around China. Book printing is born—impressions on rice paper are made with the use of carved boards.

It is during this period that on the Island of Java the famous temple Borobudur is built, resembling a tall pyramid nine tiers high and adorned with 104 Buddha statues and 1,460 bas-reliefs. A little later, the Hindu complex of Shiva temples is built in Prambanan. Art critics believe that the only comparable to these majestic structures are the great mosques of Cordoba, Damascus, and Kairouan.

The tenth century is the one of continuous warming and so barren of events that the three-kilogram (seven-pound) Atlas of World History was not able to record any notable events of that period. This is how Stefan Zweig describes these dismal hard times:

"The year 1000—the heavy, oppressive dream paralyzed the Western World. Eyes are too tired to look around; senses are too dulled to show curiosity. The human spirit is paralyzed like after a deadly disease; humankind no longer wants to know anything about the world it inhabits. And most amazing is that everything humans knew earlier is now incomprehensibly forgotten by them. They can no longer read, write, count, and even kings and emperors of the West can't put their signature on a parchment. Sciences are stagnant and have become mummies of theology; mortal's hand is no longer capable of depicting or sculpting own body. An impervious fog covered all the horizons. No one travels anymore or knows anything about other lands; people hide in their castles and cities from savage tribes invading frequently from the East. People live in tight spaces, live in the dark, live without daring—the heavy, oppressive dream paralyzed the Western World.

"Occasionally in this heavy, oppressive somnolence, a vague remembrance sparks that once the world was different—vaster, prettier, brighter, more inspired, full of events and adventures. Weren't all these countries cut through with roads, weren't the roads used for passage by Roman legions followed by lictors, protectors of order, men of the law? Didn't a man once exist by the name of Caesar who captured both Egypt and Britain; didn't triremes cross the Mediterranean Sea and reach the countries where for a long time now not a single ship dares to sail out of fear of pirates? Didn't some king Alexander once reach India—that legendary country—and return home via Persia? Weren't there any wise men in the past who knew astrology, the wise men who knew what shape the Earth has and who mastered the mystery of humankind? All this should be read about in the books. But there are no books. It would be useful to travel, to see other lands. But there are no roads. All is gone. Perhaps, all this indeed was just a dream."

But we remember that the times of climate improvement (warming) in Europe and nearly the whole world are the times of climate worsening (droughts) in Middle East, Asia Minor, and Plateau of Iran. It means that in Asia things should be fine with the culture issue. And it was indeed so. This was the time of appearance on the world cultural scene of Rudaki and Firdausi, historians Al-Tabari and Al-Masudi, geographers Ibn-Fadlan and Al-Istakhri . . . (Ibn-Fadlan was a great missionary and cultural ambassador, who went to Volga Bulgars and converted them to Islam.) And finally, the world famous Avicenna (Abu Ali Ibn Sina) and all-round scientist Abu Rayhan Biruni were also children of that time.

The climate situation became very interesting in the twelfth century. It started with a period of the next brief cooling. It ended with an era of a sharp temperature rise—the blessed time. And how do you think was the situation developing? In full compliance with the climate. The first, cold, half of the century—thirst for knowledge awakens in Europe: universities open in bunches in Bologna, Paris, and Salerno. In the second, warm, half of the century—thirst for knowledge vanishes—heaven knows where, and only one single university opens in Oxford by force of inertia gained in the first half of the century.

In the very end of the twelfth century (fig. 6), after the warming peak, in merely some twenty years temperature suddenly falls huge by half degree (-1°F). Then during a whole century it floats somewhere in the neighborhood of "-0.1°C (-0.2°F)," after which if falls further down. The number of universities in Europe is growing as if using growth hormone! While by the late twelfth century, there were only four above-mentioned universities in Europe, in the thirteenth century, there were already eighteen, in the fourteenth—forty, and in the cold fifteenth century, famous for its obscurant witch-hunt, Europe had more than sixty universities!

During the eleventh century cooling, book printing is improved in China—they start using typesetting printing instead of carved boards. The fifteenth century is China's era of great geographical discoveries. Chinese seafarers furrow the Pacific and Indian oceans. A hundred years before

—

the Portuguese, the Chinese admiral Cheng Ho completes four amazing voyages, reaching South Africa, discovering Madagascar, getting in the Red Sea, and landing in Sunda Archipelago. The admiral's expedition had sixty-two ships with the number of personnel reaching eighteen thousand. For comparison, a hundred years later Vasco da Gama reached India on four ships with the number of expedition personnel two hundred people.

The temperature curve in the fourteenth century repeats that of the twelfth century: cold in the first half and warm in the second. And it so happened that it was the first half of the century when the era of renascence took place in Japan—literature masterpieces are created, art of ikebana appears, theater and the famous tea ceremony come to being . . . And the second, warm, half of the century sees total indifference toward culture—historical quiet! As to Europe, the fifteenth century became the century of great geographical discoveries. The civilized humankind remembered once again that the Earth is a globe and discovered on this globe-shaped Earth so much new (well-forgotten old) stuff that historians start the count of the New Time from this era. Columbus, Vasco da Gama, Magellan,—these persons replaced the old decorations of the Middle Ages with a completely new decor—the decor of the New Time.

In the same dismal yet great fifteenth century, a European Johannes Gutenberg, who naturally knew nothing of the Chinese typographical experience, invents book printing. Martin Behaim makes the world's first terrestrial globe with diameter 1/2 meter (20"), which is exhibited today in the Nuremberg museum. And art creations are produced by Botticelli, Leonardo da Vinci, Raphael, Michelangelo, Giorgione, and Albrecht Durer.

Assembling and Disassembling a State in Forty-Five Seconds

Having discussed spirit, let's turn to issues of powers.

By the mid-seventh century, the Chinese Empire under Tang dynasty reached territorial maximum—North Vietnam, North Korea, Fergana Valley, East and West Turkic Khaganates—they were all under the power of the Chinese crown. The statehood of China was strong, oriental, and nomenclature-socialistic. The power was in the hands of state-class bureaucracy. Deployed all over the country was a system of state schools and standard exams that when passed gave local feudalist's son a right to become a state employee and to take bribes.

—

However, the world was turning warmer and warmer, which meant that the central power was counting its final days. In the very peak of the thaw, the country began shaking from fever of separatism. An uprising was started by a military governor of one of the provinces, whose name was An Lushan. Strives began inside the emperor's court, which ended up shamefully—by the end of the eighth century, the country's power was grabbed by a woman, the wife of the last emperor. At the same time, the country received several painful military slaps in the face from former vassal territories. During all of the ninth century, China was shaking in full compliance with temperature fluctuations, and at the peak of the warming in about 900, the Tang dynasty was overthrown in an internecine civil unrest, and the country broke up into nine pieces. During the entire tenth century, these pieces were unsuccessfully fighting one another.

Let's go back to the start of our graph (fig. 6). The paunchy era of warming seen on the graph at the extreme left is the time of the rise of the Arab Caliphate. The Caliphate "poured" the green paint of Islamic expansion over the territory that was far even from any dreams of the Roman Empire at the time of its maximum might. It seemed this superpower had everything for long-time existence—common language, common religion, common culture . . . The only problem was the weather: the country was unified only for as long as the era of bad climate lasted. Global temperature started its move down, first slowly and then swiftly; in the arid territory of the Caliphate, humidity was on the rise and people were having satiety shine in their eyes. And the Caliphate began to unglue right before people's eyes. The first one to fall off was Andalusia, then Maghreb, and then Ifriqiya (North Tunisia). Various small state entities were formed, like Emirate of Cordoba, State of Idrisids, Emirate of Aghlabids, and Imamate of Rustamids . . .

Harun al-Rashid was undertaking heroic efforts to prevent the collapse of the empire, but pieces kept falling off. The ninth century with changeable yet relatively cold climate (humid and good for that area) brought several more presents to the Caliphate—breaking off Egypt, Yemen, and Syria. By the end of the ninth century came another trouble—the empire has weakened so much that impudent Bedouins robbed the Kaaba temple in Mecca. Caliph was left with only a piece of Mesopotamia and Southwest Iraq.

That's how the earthly glory comes and goes.

Socioclimatic pendulum made its next huge stroke: at the same time as the Caliphate was falling apart, a new European empire was expanding like yeast dough (the colder—the faster). Its crystallization seed was Charles, the elder son of the King of the Franks Pippin the Short, and his small kingdom. Charles inherited from his father a worm-shaped strip of land along the Atlantic. It became the starting point of the great empire.

Charles reigned nearly half a century. And in this time, his empire spread from the English Channel to the Carpathians. Charles was so huge, talented, and magnanimous that he remained in history under the nickname "the Great," while the Pope of Rome crowned him in 800 with an imperial crown. Thus, the Western Roman Empire was in fact restored. And it did not surprise anyone: memories of the Roman Empire were still too fresh. (By the way, Charles was even called à la Roman—Carolus (or Karolus) Magnus and he spoke fluent Latin.) At the same time, the second large piece of the glorious Roman Empire—the Eastern Roman Empire—never ceased to exist and Byzantines always considered and called themselves Romans ("Romeos"). Thus, the arena of history saw a remake of the greatest of all global empires.

. . . Since Latin was mentioned, in honor of Karolus all subsequent European rulers were called kings (*in Russian, "king" is* korol *pronounced "karol" with stress on second syllable—Translator's note*) . . .

But once again, the empire did not last long—warming waves of the ninth century killed it. And then a grand warming of the tenth and eleventh centuries began. This warming, which in Asia Minor is always equivalent to worsening of climate, adds strength to Byzantine tormented by Arabs. It rose like a phoenix from its ashes and reminded the world of its former grandeur. Byzantine armies dislodged the Arabs from Asia Minor, Armenia, Syria, Crete, and Cyprus. And shortly after, they captured Serbia and Bulgaria.

However, when the period of climate inconveniences passed, being replaced by cooling that in Byzantine is equivalent to precipitation, or climate improvement, the empire ingloriously ended. In the mid-eleventh century in the battle of Manzikert, a city near the Lake Van, the troops of the emperor Romanus IV Diogenes suffered a crushing defeat from Seljuk Turks and then had to pay contributions to the Sultan Alp Arslan. The troubles happened in the west, too, where the armies of Normans dislodged Byzantines from southern Italy. That was how the fame of Byzantine waned, and this time it was for good.

Now let's go back a bit in time and move down on the map—Africa. It was here where the fragments of the Arab Caliphate were located after the collapse of this once great state. In the tenth century, when it started turning warmer and Europe began to gradually submerge into an eventless darkness, centripetal trends were prevailing in the drying-up Africa—Egypt, Syria, Palestine, and nearly all of North Africa were tied up by an iron bolt of the Fatimid dynasty—direct descendants of the Prophet's daughter. It's interesting that the centralization rigidity in the Fatimid country was such that land, all major means of production, manufactories, transport, shops, and even housing were considered the state property. It was the very real nomenclature socialism.

In the next century, in spite of the start of global cooling, amount of precipitation in Africa didn't increase and to the contrary continued to fall. That drought swept away African nomads from West Sahara, the Berbers. In search of better lands, they strolled like innumerable locust all over North Africa, captured Maghreb, jumped to the Iberian Peninsula, captured Islamic Spain, and on the way delivered crushing defeat to Castilian armies. Then they went south and captured Ghana, grabbing at the same time the lands all the way to Senegal.

The first half of the twelfth century was cold and thus humid (or climatically speaking, good for North Africa), leading to lax discipline in the Berber state. Therefore, as soon as the global temperature in the second half of the century began moving upward and the arid areas felt the drought, a Moroccan uprising occurred, bringing to power a new dynasty, the Almohads. They created a new state, a very restrictive Muslim Empire, where any religious heterodoxy was severely punished without pampering. The Almohads dislodged the Normans from all the Maghreb fortifications they occupied and paid another visit to Spain where they once again beat the hell out of the Castilians—just in order of things.

But then the dry period ended, temperature fell, humidity increased, and in the early thirteenth century, the Almohads began suffering one defeat after another. The Almohad state did not survive to mid-century ...

In the same early thirteenth century—the graph (fig. 6) shows the catastrophically fast temperature drop from its highest peak around 1,200—nomadic Mongols came to China. They were squeezed out from Mongol steppes by drought and after stubborn and long fighting conquered China. That's how the Mongol dynasty Yuan started.

It was in this era that China became the most developed nation in the world. The dynasty established diplomatic relations with Japan and countries of South and East Asia. Regular transportation of commodities was set going between China and India. Astronomy, medicine, and mathematics developed immensely. Along with Arabic numerals, Islam came to China.

In the late thirteenth century, Italian trader Marco Polo completed a travel from Venice to China. He spent there seventeen years and wrote his famous book summarizing his journey. It's interesting that Marco Polo reached China by dry land but went back by sea—along the entire Asian coast on Chinese ships, which speaks of well-developed navigation during times of Yuan dynasty. He was assigned to go to Persia with a delicate mission of escorting the great Kublai Khan's daughter destined to marry the heir to the Hulaguid empire throne. Not surprisingly, with such well-established trade routes the famous Chinese porcelain gets all the way to East Africa. But China of Yuan dynasty is famous not only for its porcelain. Chinese folklore and drama achieve great heights in that era. This was the creativity time of Guan Hanqing, Wang Shih-fu, Pei Pu, and Ma Chih-yuan . . .

The great China of Yuan did not live long: the empire collapsed when at the end of the fourteenth century, a brief, but strong warming came.

The fifteenth century was retained in reader's memory for its unusual frosts (summer snow) and flights of human spirit (Leonardo, Magellan, inquisition . . .). As to the states and their development, this century was also quite full of events. More than just full!

Entire Europe began to build actively at that time, that is building governments. Internecine wars stopped; France, Spain, and England became unified centralized states.

Northern Europe (Denmark, Norway with Iceland, and Sweden with Finland) forms at that time a unified confederation, the so-called Kalmar Union. All the countries listed in parentheses, under the keen leadership of Denmark, joined in the Kalmar Union not at all for the purpose of joint catching of squid (*in Russian, "squid" is "kalmar"—Translator's note*) in the northern Atlantic, but rather exclusively for the same reason why modern Europe united—to facilitate free trade. The Union held for over one hundred years and fell apart in the early sixteenth century (you don't even have to consult the graph—it was in the period of the next climate improvement).

In Central Europe in that same fifteenth century, the ones who prove their worth are the Hungarians who are by this time ultimately civilized. Under the rule of their king Matthias, they develop intense military activities and bring under their rule a big chunk of land extending from the Oder River to the occasionally freezing up Adriatic and from Bohemia to Transylvania.

In East Europe at that time, a powerful Poland-Lithuania state is formed, extending from sea to shining sea—from the Baltic Sea to the Black Sea.

In Russia, Ivan III, "the Great," brings all Russian lands under his rule. Karl Marx wrote, "While in Ivan's early reign Europe hardly noticed the existence of Muscovy squeezed somewhere between Tatars and Lithuanians, it was astonished by the sudden appearance on its eastern borders of a huge state, and Sultan Bayazid himself, having the entire Europe at his feet, heard for the first time the Muscovite's supercilious speeches."

In the south, the Osman Empire expanded, occupying the entire Asia Minor and sitting as a heavy bulky beast over North Africa. It came to Europe where European capitals fell one after another: Sofia, Belgrade, Esztergom, while Vienna was under a heavy siege.

In South America, the Inca Empire was swelling, extending from modern-day Ecuador to Chile.

In West Africa in the fifteenth century, the Gao (Songhai) Empire is gaining strength—the most powerful state in West Africa. It extends from the Senegal riverhead in the west to the Ayr Mountains in the east and all the way to Central Sahara in the north. When looking at the map of modern-day Africa and trying to outline the Gao Empire, it includes Senegal, Mali, Niger, Nigeria, and several other smaller African countries.

In Southeast Asia, the great Vietnam swelled like a toad, occupying the entire eastern Indo-China.

We Are Not Locals . . .

Let's once again glance at the Middle Ages in order to look this time at the boiling history from a different angle—how climate worsening affects people's bellicosity and dislodges savage nomads and other conquerors from their longtime homes. And we shall start again with Arabs. You have already guessed, probably, why . . . Because on our graph (fig. 6), the first bend of the temperature curve is upward.

Global warming is equivalent to worsening of the living conditions in the arid area. That is, in those places where Arabs live. Agriculture failed so sharply then in the East, that there was almost nothing to eat.

My university schoolmate Ben once asked me, "Let's say, a terrible crisis happens. What are you going to do when all the money and food in the refrigerator are gone?"

"I'll find either a full-time or a part-time job."

"And what if you'll find none?"

"I'll start selling my belongings."

"And when all would be gone?"

"I'll go begging."

"But nobody will give alms, because famine will have set! What are you going to do then?"

"I'll start robbing and killing . . ."

"But why wait until you become emaciated and weak and have sold all your belongings including the ax that could be useful for the killings, when you can foresee everything in advance and start robbing and killing while you're in good health and full of creative energy?"

He was right, of course. But most people are created in such a way that they try to resist the war until the very uttermost. And only when it becomes totally unbearable, they go robbing and killing. But at that point, their despair has no limits.

In the early and mid-seventh century, Arabia was shaken by serial catastrophes. Several powerful earthquakes and eruptions of volcanoes, extinct by now, along with the drought have worsened the life of local population to such degree that it went robbing and killing in all four directions—north, south, east, and west.

In the west, the ringleader of one of the gangs, Tariq ibn Ziyad, crosses over the Gibraltar (which name comes from Jabal al Tariq, meaning "mountain of Tariq," or Gibr Tariq—"Rock of Tariq") and captures Cordoba and Lisbon. At the same time, Arabs invade India and capture the city of Multan. If you look at the map, you'll see that Cordoba and Lisbon are almost 7,000 kilometers (4,500 miles) apart, leaving an impression of the unimaginable force of the Arab onslaught.

A little later, the expanded empire in the east becomes contiguous to China of Tang dynasty. The Chinese armies are defeated in 751 at the Talas River. By this time, the era of drought is gradually ending, thus ending the robbing-missionary rush of nomads.

The next cooling leads this time to climate worsening in Europe and serves as an awl in the butt for European integration under the leadership of Charles the Great.

After the cold eighth century came the warm tenth to eleventh centuries, when Europe, satiated with sufficiency, does nothing but defending itself from the savage nomadic hordes that are driven away by drought from the steppes to places where they could gorge. What all these Magyars, Saracens, Khazars, Polovtsians, and Pechenegs do is only robbing and killing, robbing and killing . . .

And not only they. The era of ninth to eleventh centuries was the era of Vikings. That's who vexed Europe not childishly! In the warm era of late seventh to early eighth centuries, so many Vikings populated northern areas that the start of cooling and associated lack of food squeezed out this mean and cruel people toward Europe, for they had no other way to go. A saw was even born in Europe, "Save us, oh Lord, from devil and Normans."

Here's a list of all the most remarkable acts of these northern barbarians, starting with the minor:

793—attack on a monastery in Lindisfarne,
799—capture of Nantes,
841—capture of Ruan,
845—capture of Hamburg,
844—capture of Lisbon and Seville,
846—siege of Rome,
859—landing on the Balearic Islands, in Catalonia, and in southern France.

During all of the ninth century, the Normans were intensively resettling from their god-forsaken places, while establishing colonies in England and Ireland and sailing to the Faeroes and Iceland. Detachments of Vikings went as far as Byzantine, the Volga, and even Baghdad Caliphate.

All this was happening in the cold era. But as soon as it got warmer in the tenth century, the former activity of Normans immediately evaporated somewhere, like the dew under the bright sun. And gradually they started being beaten—in England. The Norman political leader Canute the Great managed to get the reins of power into his hands, but the global warming did not allow the Normans to do what they earlier managed easily: the reins barely drawn by the hands of Canute the Great were slackened right after his death. The only thing left was to wait for the new wave of cooling.

The temperature plummeted in the eleventh century. And heartened-up Normans captured Sicily and created a powerful kingdom, not only dictating its will to the pope and to Byzantine but also being able to get the better of the pirates of Maghreb who were the sole masters of the Mediterranean. This is what the Arab historian Ibn Khaldun wrote about total pirate predominance of that time: "Christians could not lower on water even a plain board." Christians could not get the better of the bandits, but Normans did.

It was the Sicilian kingdom of Normans, which thanks to dynastic marriages produced the super-cultured Friedrich II, whom we talked about earlier in this Part. It was the same Friedrich II, who became the ruler of the third incarnation of the Roman Empire—the Holy Roman Empire of the German nation.

And this is where we smoothly and imperceptibly turn to the era of the Crusades.

The Crusades toward the Lord's coffin became fashionable after the famous speech of the Roman Pope Urban II, which he made at the Council of Clermont on November 26, 1095. The pope said in particular, "The land you inhabit became overcrowded due to your multiplicity. It is not abundant with richness and hardly provides bread to those who cultivate it. Following from this is that you bite one another and fight one another. So now your hatred may stop, your animosity may calm down, and internecine feuds may cease. Take your path toward the sacred coffin, extort the land from the impious people, and override it. Those who are unhappy and poor over here will become rich over there."

... So that was in fact the reason for the Crusade hikes. Europe became overly populated, and its feeding base was sharply reduced ...

Why the feeding trough became depleted is quite clear: about forty years before the pope's speech, the global temperature was falling at a rate of high-speed train. It dropped 0.2°C (0.4°F) and kept inevitably falling. That temperature drop was enough to make "land ... overcrowded." With that one speech, what the artful Pope of Rome achieved was that tens of thousands of extra mouths quickly collected their belongings and left far away from the feeding trough.

As Klimenko himself notes in one of his works, "In the confrontation between West and East, the strategic advantage was always on the side of the former in the cold times and of the latter in the warm droughty eras, that is those were always stronger who were at that time in the worst climatic conditions."

The same was true for Crusades. The First Crusade, taking place in cold era, was successful: Jerusalem was captured, and several Christian states were formed in Palestine and Syria.

When it warmed up (second half of the twelfth century), those Christian states were defeated by aboriginals, and the Second and Third Crusades ended with failures.

The start of the thirteenth century was a whole half-degree (1°F) cooler than the end of the twelfth century. And Christians had a better luck this time. They managed to recapture from the Moors almost all of the continental Spain and the Balearic Islands. (The Moors final ousting happened in the fifteenth century—also a cold one.)

It becomes even funny—in the cold decades of the century, Muslims suffer defeats: they lose Cordoba and the Baleares in 1236 and Cadiz in 1262. And during the heat wave in mid-century, King of France Louis IX is taken prisoner with his army in Egypt (1249). Let's assume these were coincidences. But slight warming at the end of the thirteenth century brings Christians no luck either—knights of the Order of St. John leave the city of Acre, the last stronghold of Christianity in the Middle East. Without going too far away from the thirteenth century, let's see who and wherefrom was ousted by the global cooling that started twenty years before that century—in about 1180. As you can see (if you're not lazy to rustle some pages), it turned cold very quickly and very sharply. In fifteen to twenty years, the temperature dropped half degree (1°F). We discussed this period already. The Mongols . . .

The cooling in the Mongol steppes was accompanied by a drought, which fact is supported by frequent dust storms at that time. So, the Mongols ran away from this double trouble, robbing and destroying countries and people on their way.

Gumilyov was the first to point at the drying up of steppes as the reason for Mongol invasions. However, recent paleoclimatic reconstructions show that the picture is not as simple as the passionate old man thought: the Eurasian steppe zone, extending from Hungarian plains to China, dries up and moistens quite unevenly—a drought could be in one place and lots of precipitation in another. The atmosphere—ah, hell with it . . .

For example, the steppes, wherefrom at the turn of the fourteenth to fifteenth centuries, the nomads of Tamerlane went to crush the Osmanis, were dryer than the Osman Anatolia. And when Anatolia had been conquered by the Osmanis earlier, their lands were drier than Asia Minor. But all these are just minor details that we shall not digress into and rather get to the major stuff.

PART 6

Future Starts Yesterday

The worlds have between them no difference,
As well as they have no sum total.
The steppe is so cold that it stiffens
And freezes to death common mortal.

The snow makes all shiver and tremble,
Half Europe is equally chilling,
And demons come out of the temple
And play in the snowdrifts like children.

Alexander Anashkin

In Sevastopol inlet and Georgian land, all is abundant.
And Turkish land is plentiful. In the land of West Ukraine,
too, all food products are in abundance and cheap. All is also
abundant in Podolsk land. May be the Russian land blessed
by the Savior.

Afanasy Nikitin (1466-1472)

The Russian scientist Mr. Klimenko periodically travels abroad to earn some money. Russia so far feeds its scientists poorly. There was a time when it was so bad that Klimenko had to finance his Laboratory of Global Energy Problems and pay employee salaries out of his own pocket. That is, out of the money he earned in Europe. And I don't quite understand why he returned to Russia in the first place. Each has his own oddities . . . Have you paid attention to the name of the laboratory—". . . Global Energy Problems"? Energy is very closely related to climate, and we'll surely talk about it later. But we shall start, of course, with climate.

"So, national psychology is defined by climate . . ."

"But by what else? By climate and geography."

"But Vladimir Viktorovich, you'll agree that the essence of progress is that humans move further away from being part of nature, become more independent of nature. Thus, our dependence on climate decreases, but our dependence on technosphere increases."

"That's correct. But where are you leading to?"

"I mean that the carrier of national psychology is to a great extent a rural citizen, a peasant, but not a city dweller. City dwellers are less dependent on climate. I wrote in one of my books that the difference between a Parisian and a Muscovite is less than between a Muscovite and a village dweller somewhere near Tambov."

"You know, the way of life and economy change instantaneously—on a history scale. While psychology changes in centuries."

"I don't think so. Or even quite the opposite—I categorically disagree! Two-three generations developed in the conditions of a large industrial and even more so postindustrial city—and the national psychology almost entirely is erased. And you see a person with no nationality—a city dweller. All that's left from his nationality is language and some minor stuff, shell . . . And sometimes people change even faster. For example, psychotype of a Soviet citizen and of a modern-day Russian differ quite significantly."

"I don't think so. I judge by myself. I used to be in a village only when visiting my grandparents during my vacations, but wherefrom do I have the so-called egalitarian aspirations, or yearning for fairness and equality? I like such countries as Finland, Norway, Iceland, New Zealand, South Argentina, and Chile, where there are no rich or poor. They have no millionaires and no paupers."

"Hum . . . But wherefrom do I get the opposite—the sharp liberal rejection of socialism or egalitarianism? We grew up in the same country . . . It's simply deviations on the level of person's psychology, I think . . . By the way, regarding South Argentina and Chile . . . Is the situation in countries of Latin America the same as in Scandinavian countries known for their socialism? I know that Latin America is characterized by its extreme polarity—some live in villas, while others in slums."

"I'll respond in the order the questions were asked. And it's all about climate's effect on psychology. I've been to all northern countries of the world. And I've noticed: the further north you go, the stronger egalitarianism yearning is developed in people—so no one would stick their head out. About six years ago, I took part in a conference dedicated to changes in Arctic climate. It took place in April in trans-polar Norway. The city of Troms (69.5° northern latitude), which hosted the conference, was buried under snowdrifts. The streets were literally dug through like tunnels—on both sides of the road were vertical four-meter (13-ft) high snow walls. So, I was staying there with a Norway millionaire. He's a ship-owner, which should tell you accordingly the financial scale of this person. But he owns a small wooden house on the edge of a fiord. His car is a twelve-year-old Ford that cost no more than $20 grand, even new. I asked him why he was driving a car like that. He answered that he was quite happy with it. And this situation, this kind of psychology of the northern rich is a norm.

"As to the shameful luxury and huge difference in living standards—they could be found only somewhere in Africa, Russia, Arab nations, and as you correctly noted, in Latin America. In South America, the contrast is simply crying. They have villas with huge swimming pools and even football (i.e., soccer) fields (everyone there is crazy about football [i.e., soccer]). About 2-3 percent of the population lives like that. The rest live in cardboard boxes, shacks, and barns, made of rusty roofing iron and palm leaves."

"In Ecuador, I was a witness to a remarkable incident I saw from a bus window. A high fence encloses a rich man's estate. There are security cameras, and on the sidewalk along the fence walks a guard in black uniform—a sturdy fellow . . . A man walks on the sidewalk—frail and small. All of a sudden, the guard raises his club and hits the passer-by—for daring to walk along the fence of a decent citizen. The passer-by falls down, then gets up, starts apologizing, and then goes to the other side of the street. That's Latin America. And it's not very far from the equator. But situation changes closer to poles. It's altogether different in the very south of Argentina—in Ushuaia, where climate is severely cold. You don't see any paupers or beggars. Is it maybe because they simply can't survive there and freeze to death? It's like two different countries—the cold South Argentina and the Argentina closer to the equator. The cold South Argentina reminds of the Scandinavian countries. Perhaps, severe climate creates the kind of human psychology when it's indecent to be rich."

"Is it for this reason that everything is so good in the northern countries? In Iceland—an awfully rich country where summer temperature rarely rises above 15°C (60°F) with practically no trees but only grass, moss, and wet rocks—they grow bananas and pineapples, and not as decorative plants but commercially! In giant hothouses the size of ten football fields, a real rain forest grows under a roof. The entire city of Reykjavik is buried in flowers. And it's under leaden Arctic sky! Although Iceland has neither coal, nor oil, nor gas, but it has great well-being. I've been to Reykjavik, in the block of the rich. Their modest houses cannot compare with the villas of South-American or Russian nouveaux riches. And this modesty is equally important for both those who live in these modest homes and those who walk around them. I personally am frustrated with what I see on the river shores when sailing on board a ship from Moscow."

"So you're a socialist, old chap!"

"You know, when I was in Reykjavik (I was there with some Germans), the Germans when seeing this northern miracle exclaimed more than once that the Icelanders were living under Communism. The Icelanders (as well as the Swedes and the Norwegians) call their social structure "socialism," which is only natural because their ruling parties more often than not are Social Democrats. We just had different interpretation of the works by Proudhon, Marx, Bakunin, Kropotkin, Engels, et al . . ."

"Hum. In a peasant country, such as Russia, Kampuchea (Cambodia), Korea, or China, there could be no other brand of socialism except the one with a peasant face. But as soon as village was facing its end, the village socialism ended, too . . ."

CHAPTER 1

Impossible Way of Life

Since we've mentioned it, it's now time to talk about the enigmatic Russian soul, how the Russian climate affects it, and in general—why are we the way we are? What are the physical reasons for this generic curse of the nation, this misfortune called Russian character? But we'll start with two stories of characters . . .

My friend's father bought a dacha (summer country house) in the true Russian boondocks. In the heart of Russia (I apologize for the grandiloquence)—the Upper Volga. It was not even a dacha but rather a plot of land in a village, with remnants of a house on that plot. And he began to build. It was remarkable and pleasant to see how the locals were all helping the newcomer. And by the way, they were doing it free. They were leaving their own gray, old, shabby huts with small doors, and were sawing, carrying . . .

. . . Kind, responsive people . . .

But the Muscovite, to his own misfortune, built a big two-storey house full of light. With huge doors that allowed one to enter without bowing. The house was good. It shined in the village like a golden tooth crown amid the line of rotten used-up teeth. And soon the same people, who helped their neighbor build the house, burned the house down. So it would not stand out.

. . . Envious, mean brutes . . .

Now the second story. In a Moscow tourist office, a married couple is selecting a hotel in Turkey.

"Are there many Russians there?"—Asks the wife, an ethnic Russian.

"No, don't worry, not many at all."—The clerk soothes her, also a Russian.

"That's good, because I'm tired of them; it's even impossible to relax. And they behave like pigs."—The husband nods in agreement, a Russian, too.

Living conditions form a person's mentality. Scientists and folk healers like to repeat, "A human is that what he eats." And I would add "... and how he obtains his food."

The process of mass transfer from village to city (civilizing) was completed in Russia relatively recently. What the retired city dwellers sing when they gather? Drawling village songs. The middle-aged generation has still in its memory the word-combination "writer-villager." And the old Soviet colored movies of early 1970s are still running (although more and more rarely), showing the tough rooting of villagers into a city.

True, today Russia is an urban country with respect to its mentality and culture. But yesterday it was rural. Well, what can I say, goddamn it, if I'm myself only a first-generation Muscovite! And my father (I'm sorry for the striptease) saw a "real" steam engine for the first time only at the age of sixteen! Centuries of peasant labor formed the mentality of the nation consisting 90 percent of peasants. And it had not enough time to change in the last 150 years of "urbanization." The heavy air of rotting bast shoes has not yet completely evaporated from us.

So what is it in the Russian rural way of life that made Russians the Russians—lazy, unreliable, sloppy, careless, inept, always counting on an off-chance? As to the details of the Russian way of life, I once consulted elaborately one of the best specialists in this field, professor of History School of Moscow State University Leonid Milov, who studied peasant way of life throughout his career. Who if not he would know about it?

Milov and I were sitting on a bench in one of Moscow parks, and red-faced passers-by in sweat pants looked forebodingly askance at the conspicuous red light of the Dictaphone I was holding. Somewhere close by, behind the bushes, two drunks were fighting. The third one, rather than separating them, was puking strainedly, his belly roar filling the neighborhood. It was a quiet evening, typical for that time of the year.

"Russia is a very cold country with poor soils, which is why only people of this kind live here and none other. In Europe, agricultural period lasts ten months, while in Russia only five."—Milov was telling sadly. "The difference is twofold. In Europe, they don't work in the fields only in December and January. In November, for instance, winter wheat could be sowed, and English agronomists knew it back in the eighteenth century. In February, other work is done. So if you do the calculations, it turns out that a Russian peasant has just one hundred days for all the land works except grain threshing. And thirty days for haymaking. What do we get? It's that he torments himself and hardly manages to do all the work. The head of family of four (single-working peasant) manages physically to plow two and a half dessiatinas (6.75 acres), while in Europe it's twice as much.

"The fact that land-noncultivating period lasts in Russia seven months was reported in state documents back in the eighteenth century. The problem was understood . . . Average yield with those tools was "self-three." That is, one grain produced three, 12 poods produced 36 (1 pood = 36 lb.). Minus one pood for seeds, result is 24 poods (864 lb.) of net yield from a dessiatina (2.7 acres), or 320 lb./acre. Two and a half dessiatinas (6.75 acres) give a net yield of 60 poods (2,160 lb.). And that's for a family of four. Considering that women and children eat less, a family of four equals 2.8 adult men. The annual per capita consumption norm was 24 poods (864 lb.). Thus, the family would need about 70 poods (2,520 lb.), while only 60 poods (2,160 lb.) available. Also, it's necessary to subtract some for the cattle—oats for the horse and grain added to cow fodder. And instead of the biological norm of 24 poods (864 lb.), an average Russian had 12-15-16 poods (432-540-576 lb.). It equals to 1,500 calories consumed a day instead of 3,000 needed for the body.

"And that was the average Russia—the country where grain was always in short supply. Where life was always at the limits of possible. Perpetual struggle, perpetual fear of hunger. And at the same time, working heavily to the ground with involvement of women, children, and elderly . . . And was it possible to expand the arable lands? Yes, if working haphazardly with off-chance reliance. And that's how they worked. While in England the land is tilled four to six times, bringing it to the 'downy' state, in Russia the land cultivation is ugly up to this day. And although the equipment has changed, with tractors used in both Europe and Russia, but tilling time ratio remains the same and result is the same, too: in Europe, it's impossible to find even a tiny lump on the cultivated land, while in Russia huge cobbles lie in the field. True, labor productivity in agriculture has increased forty to fifty times when compared to the eighteenth century. But the nature remains the same! For this reason, for the reason of climate, the prime cost of Russian agricultural products will always be higher than in the West.

"Have you seen the movie *The Chairman*? Do you recall a heart-rending scene when women use ropes to keep the cow upright and prevent the weakened animal from falling? That scene is typical for Russia. By spring, cows and horses can hardly stand. With all the enormous vastness, fields, coppices, meadows—and yet, peasants don't have enough hay. Why? Because when the grass is full of vitamins and needs to be stocked and stocked, peasants have no time for it. Haymaking time always started at Peter and Paul holiday—June 29, by old-style Julian calendar (July 11 by Gregorian calendar in the nineteenth century, July 12 in twentieth to twenty-first) and lasted one month. And in August, and sometimes even starting July 20, it was already time to reap the ripened rye.

"Thus, despite that during the haymaking period the entire village from kids to seniors went mowing and peasants lived encamping in the fields, with the mowing tools they had then they still could not mow enough hay. The period of keeping cattle stalled in Russia is 180 to 212 days, or six to seven months. An average single-peasant household (four persons) had two cows, one or two horses for plowing, two sheep, one pig, and five to eight chickens. Goats were infrequent. And those numbers could differ from district (county) to district. For example, in Tver province a peasant in one district had three sheep, while in a nearby district it was three to four pigs. But in calculating the equivalent averages, it amounts to six head of cattle. In accordance with the norms for eighteenth century, they would need to stock 620 poods (11.2 short tons) of hay. And the best a peasant could do with his family is to stock 300 poods (5.4 short tons), or less than a half of the norm. And it was always like that.

"So what could they do? Cattle received straw that is low-calorie and has no vitamins at all. But they had not enough straw, too! Pigs and cows were fed horse dung strewed with bran. Perpetual headache for heads of collective farms and Russian landowners was the chronic fodder shortage. By spring, animals would literally fall and had to be suspended. And there was not much dung from those animals, too, to say nothing of the milk; in some provinces, cows were kept not for milk, which they practically did not produce, but for the sole purpose of dung that was also in short supply for obvious reasons. It took years to accumulate dung!

"Russian cattle had extremely poor quality. And all attempts by landowners and educated people from government to import high quality breeds from Europe into Russia always ended up the same—western breeds quickly degenerated and became practically undistinguishable from the poor-quality Russian animals.

"According to all the laws of the three-field system of crop rotation, the land has to be fertilized once every three years. But in real life, peasants fertilized the land approximately once every nine years. They even had a phrase: 'good land remembers dung for nine years.'" But there were places in Russia—even in the early twentieth century, where land was fertilized once every twelve, fifteen, or eighteen years. And in Vyatka province, for instance, it was once every twenty years! What kind of crop yield could be expected?

"If you thought, 'But then our peasants were relaxing for seven months a year and simply idling away their time!' you were deeply mistaken. They had more than enough work to do in winter, too. Here's an example. Because of their permanent poverty, Russian peasants, unlike the European ones, did not wear boots. In order to have boots for the entire family of four, the peasant would have to sell three quarters of all his grain, which was unrealistic. Thus, the boots were simply unaffordable. Russia wore bast shoes. In a year, a peasant wore out fifty to sixty pairs of bast shoes. Multiply it for the entire family. The bast shoes were made, naturally, only in winter, since in summer there was no time. Next . . . The peasant could not buy fabric at the farmers market. Or rather, he could, but only as a rare and luxury present—but only to his wife and never to his daughter. But they had to wear something. Thus, in winter women spun and weaved. Plus the made belts, harness, saddles . . . Stockpiled lumber for firewood . . . By the way, up until the late eighteenth century, there were no saws in Russia, and trees were felled using axes. And they needed huge amounts of firewood, about 20 cubic meters (700 cubic feet), because the stoves were far from perfect and the huts had no ceilings (ceilings as additional heat insulators began to appear only in the second half of the eighteenth century)."

. . . Let's interrupt Milov to say that the seventeenth to eighteenth centuries were the coldest time of the so-called Small Ice Age. It was even colder then than in the horrible fifteenth century, which is probably still in the reader's mind . . .

"In summer, Russian peasant would get up before 3:00 to 4:00 a.m. and go to the farmyard—to bring fodder and clean away the dung—and then work in the field until lunch. After lunch, he would have a nap for an hour to hour and a half. Men would retire for the day before 11:00 p.m., while women would go to bed later, being busy with needlework. In winter, the daily routine was practically the same, except going to bed an hour earlier."

. . . Now tell me, is it possible to live this way? . . .

The life of Russian peasant was not much different from that of primitive Neolithic savage. Except, it was worse, perhaps . . . What was a Russian hut, for instance? A low one-room structure covered with straw. We mentioned already the lack of a ceiling. The floor was often earthen. The entrance door was rarely higher than one meter (40"), sometimes only half meter (20")! A typical hut was heated "in black mode." This strange structure had no windows. The smoke from the stove got out via special tiny openings size of half a log. As to linen and even mattresses and featherbeds, Russian peasants for a long time had no idea about them and slept on sackcloth and straw. In one room, eight to ten people slept side by side on benches and planking above the stove. Animals were also here—chicken, pigs, calves . . . Foreign travelers were shocked seeing heads, feet and hands hanging down from the planking. "I had an impression they could any minute fall down to the floor," wrote Cox, the researcher of the Russian life.

Peasants stoked the stove starting in the morning. By three-four o'clock in the afternoon, it heated up a lot and the entire evening the house was extremely hot. Occasionally in the middle of the night, trying to escape the unbearable stuffiness, men jumped out into the winter freeze, all steamed and sweaty with bare chests—to cool down. And this caused numerous sicknesses and colds, often lethal. But by morning, the hut cooled down so much that sleeping men's beards froze to the planking. And since the hut was heated "in black mode," a long black fringe of soot was hanging everywhere.

And the smell! Such miasmas bloomed in a nonventilated lodging (keeping warm) that the unprepared persons became giddy. You might recall that Kharms had Pushkin holding his nose when Russian bumpkins would pass by. "It ain't yet bad, sir . . ." In essence, the country was divided into two human "subspecies"—the cultured European-educated aristocracy using porcelain dishes and discussing Ovid's poetry, and the absolutely gray, downtrodden, half-animal, superstitious mass living at the limits of possible like animals and far below the poverty level. It's obvious that these "subspecies" not just did not understand but were unable to understand each other, for there was an abyss between them. Often they even spoke different languages—one spoke French, while the other spoke Russian. Two countries in one . . . Elois and Morlocks.

When Peter I, "the Great," started his reforms, Russia had 6 percent nonpeasant population. Only six! Because the half-starving Russian peasantry simply could not feed a greater number of dependants, given local climate. And these 6 percent produced monks, nobles, military, officials, science . . . An amazingly ineffective country!

The well-being of the elite was not just strikingly but catastrophically different from the well-being of 94 percent of the population. While the "black" peasants ate oilcake and orache and collected aegopodium (the first grass with tiny flowers) in spring, the Russian nobility all year round ate watermelons, plums, lemons, oranges, and even pineapples. In order to grow tropical fruit in glass hothouses, complex systems were developed for subsoil soil heating. And it was despite the high cost of hothouse glass needed in immeasurable amounts.

From the viewpoint of an ordinary Russian, officials and city bosses are not just few in numbers and inaccessible. They are incomprehensible, as if living on a different planet. The bosses—they are like even not people, more like celestial beings. One can scold them—just like occasionally one can blaspheme, but if a celestial being condescends to one personally—then "Dear Father!"

I can never forget an episode taped by a hidden camera back in the Yeltsin era. An imposing man with a cell phone in his hand approaches an ordinary common Russian in a street. He says that he is the president's representative and asks what he, the ordinary Russian, thinks of the popularly elected president. The Russian, naturally, starts sputtering, waving his hands, and cursing. Obviously, he is not doing well! It seems if he were to see the president right now, he would just rip him to pieces. The imposing man, having listened attentively, dials some number on his cell phone, and gives the phone to his interlocutor.

"You're going to speak now to Boris Nikolayevich Yeltsin. Tell him all you think."

"Hello, my fellow Russian," comes out of the cell phone into the ear of an ordinary common citizen in an inimitable president's voice.

And a miracle happens. When the president asks how he is doing, the Russian suddenly answers:

"All is fine, Boris Nikolayevich!"

Then he fervidly wishes the president good health, turns over the phone to the owner, and continues his walk in the street. His face shows enlightenment.

Torporific daily labor, which brought no substantial fruit nor promised any perspectives, dark hopeless life, life itself on the limits of perpetual starvation, total dependence on weather could not have affected the formation of the Russian psychotype.

No matter how much you work, everything is in the hands of God; if he favors you—he'll provide, it he doesn't—you'll croak. Whether you work or not—almost nothing depends on you. Hence, the Russians have in them the perpetual dependence on the "decisions above." Hence, they are superstitious up to obscurantism and perpetually counting on the off-chance. Up to this day, the chief gods of Russians after Christ are the Great God Off-chance and his brother Perchance.

The entire living time of a Russian from the early childhood is spent, besides sleep, on simple survival. Pregnant women bend down working in the field until labor and deliver right there. It's not for nothing the words *harvesting* and *suffering* have the same root in Russian (*which is* suffer—*Translator's note*) . . . People living in the perpetual state of extreme and having half of their newborns die, no longer value the life of anyone else or even their own. And it's not they but the God who is in charge of it anyway.

Hence, the attitude toward children has an utterly consumerist nature. Children are something to help around the house. Hence, addressing beloved offspring: "To kill you would be not enough!"

When my friend Alex Torgashev flew in to me from Chicago, having lived for three years in the United States and getting unused to the Russian ways, he was simply shocked when hearing in a Russian airport how a Russian mother was yelling at her three-year-old daughter who soiled her dress: "I'll knife you!" My friend was shocked not just by the situation itself, but by the details thought through in mother's imagination of how she was going to deprive her child of life—"knife."

In Russia, children are produced not for their own sake, but "for someone to be around to give me a glass of water in my old age." "Children are our riches"—this is the worst and the most consumerist slogan invented by the Soviet authorities, as if it was borrowed from the eighteenth-century peasant Russia. At that time, children were really considered riches because they could be used to do work from age seven. Boys up to the age fifteen would work at half strength and after sixteen at full strength, meaning like a man. Teenagers were riches. Small children were a burden, extra mouths to feed. They died like flies, and nobody seemed to care—women will give birth to more! Hence, the saying: "Oh Lord, give us cattle that breed with ease and children that decease."

Europe was afraid of Russian bayonet attacks, because the Russian soldier-peasant did not value his life. His life was like real hell, compared to which the death would not be the worst option. Another Russian saying goes, "In public, even death is honorable."

"Public" here meant the peasant community.

There's an opinion that the Stalinist collective farms (kolkhoz) were tolerated for the only reason that they were in absolute compliance with the popular spirit and within the flow of former way of life. Oh yes, I mean this goddamned communality. The entire Russian peasant psychology is the psychology of collectivism. On the one hand, it's good—everyone should help one another. But on the other hand, communality is intolerant to the upstarts—the persons who stand out for something (intellect, richness, appearance) . . .

Without this collectivist psychology that impedes development of capitalist relations (whose essence is in greater atomization, or individualization of the society), the Russian peasantry could simply not survive. There was simply no way for a lone peasant to survive under conditions of plow-time deficiency when "a day feeds the year." If you were sick for ten to twenty days and did not plow the land, your family is doomed to die from starvation. If the house was burnt down or the horse died—who would help? The commune. And when the land became completely infertile and stopped bearing fruit, all peasants worked together like one community to do the clearing—the forest was removed to make way for arable land and each worker got a parcel of it. Thus, the peasantry in Russia simply could not exist as class without the help of the commune.

Commune is a terrible formation that traumatizes national mentality. In people's minds, it overcame the agrarian era and entered the industrial one. There was a children's poem in a Bolshevik era: "From his place of work my daddy brought a real and shiny saw!" Why from place of work and not from store? Why *brought* and not *stole*? The answer is all the same. "Everything is national—everything is mine!" *(Quote from a popular Soviet song—Translator's note.)* No respect for private property. Communal-socialistic concentration camp ...

Instructions from the mid-eighteenth century on managing landowner's estate noted: "Laziness, deception, lies, and theft seem to be inherited by them (peasants—A.N.) They deceive their master by pretending to be sick, old, poor, breathing falsely, and by being lazy in work. They steal what was made by joint labor, they don' want to do anything needed for safekeeping—clean up, cleanse, lubricate, wash, dry, repair ... Those assigned to be leaders have no limits in spending money or bread. They strongly dislike saving for future and as if purposely try to bring things to ruin. And they don't supervise those responsible for correct and timely work. They are silent and cover those who swindle—for friendship and honors. And they attack simple-hearted and kind people, push and chase them away. They are never thankful for kindness in awarding them with bread, money, clothing, cattle, or freedom, and instead of gratitude become rude, angry, and cunning."

Unpretentiousness and long-suffering, minimization of level of their needs ("anything but war"), disregard for others and at the same time extreme dependence on them, readiness to help and dark envy, emotional openness and cordiality that could instantly change to hatred—this is just an incomplete list of Russian person's qualities inherited from our unfortunate ancestors. And Russia enters the postindustrial twenty-first century and the informational civilization with quite a significant part of its population whose conscience is not even industrial but occasionally pure peasant or patriarchal. And if we want to survive in the new world, we have to very thoroughly, literally drop by drop, squeeze out the Russians out of ourselves. And become simply humans.

CHAPTER 2

Very Vast Is My Beloved Country

(The chapter title is the first line of a popular Soviet patriotic song—Translator's note.)

The Small Ice Age, which started on our planet in the fourteenth century and ended in the mid-eighteenth century, has turned out for Russia the way it ought to have turned out in full compliance with theory—in empery. Below is the graph (fig. 7) of Russia growth in that period. The graph is so revealing that it speaks for itself without the need for any special comments.

The start of the cooling corresponds to the start of unrestrained growth of the empire. The end of the Small Ice Age and start of global warming—we see slower growth, all sorts of twitching, and territorial losses. In other words, collapse of empires in the era of global warming is the climatic determinant, and one of the world's last empires, the Soviet Empire, was not able to withstand it. Before that, in the early twentieth century, the Chin Empire collapsed (China lost parts of East Turkestan and Mongolia), and the British Empire followed suit in the mid-century. In the century's end, Yugoslav micro-empire collapsed, and the United States is about to crack.

Russia originated in the mid-Russian forests. To the north of the forest belt is tundra where civilization simply can't be and only primitive ethnic groups can exist at the Stone Age level. To the south of the forests are steppes and forest-steppes. Steppes and forest-steppes are, of course, better "suited" for agriculture than the forest zone. And not as much because there is no need to fell or clear anything but rather due to the quality of soil. In Russian forest zone, podzol soils are predominant—they are meager and require deep plowing. Also, plenty of swamps, clay, and sand are here.

As to the steppe . . . It's made by the nature itself for the kingdom of herbaceous plants. Actually, they do reign here and that's why it is the steppe. The famous black earths! Humus is the product of decomposition of grass and other small plants. The thickness of black earth is 1.5-3.0 meters (5-10 feet). But good places always attract blockheads. Nomadic cattlemen ruled the steppes, and they disliked farmers. Thus, initially farmers did not dare to show up there out of their forests.

Russians were also unlucky with respect to precipitation distribution. For example, not too many people know that the annual amount of precipitation in the Moscow area is only tiny bit higher than on the edges of the Sahara—550 mm (22"). For comparison, Germany has 1,000 mm (40"), England has 1,500 mm (60"), and even African Tunisia has more—700 mm (28"). So what kind of agriculture could there be in Russia's middle belt? Moscow is not a desert yet only because the sun is not as hot here and thus the evaporation is lower. And besides, right after the water falls, all of it gets into the "vertical pumps"—the trees.

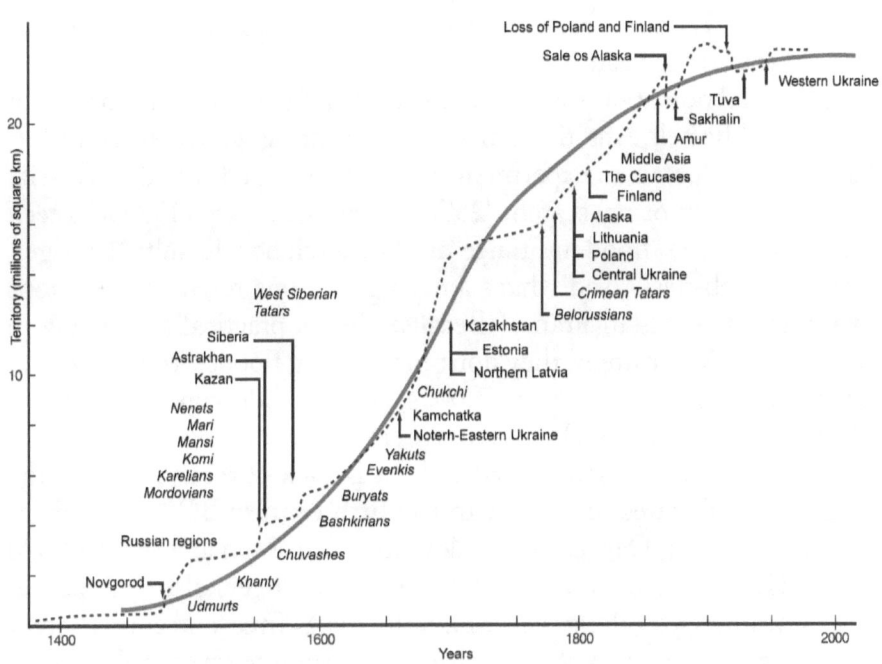

Fig. 7. Inclusion of lands and nationalities into the
State of Muscovy, Russian Empire, and the USSR.

Even in the Baltic area, they have better soils and more rains. But Central Russia is a scary zone under evil spell, where poorness of soil is aggravated by lack of moisture. And that's not all! Another feature of Russia is that precipitation falls here mostly in the second half of summer. In Moscow province, for instance, up to one quarter of all precipitation falls in July and August. In other words, a slight shift in precipitation regime can easily lead to a drought in spring and June (when young growth shoots), followed by downpours in autumn (when it's time to harvest). West Europe is luckier—precipitation distribution there is better balanced round the year.

Autumnal downpours are generally quite a frequent phenomenon in Russian middle belt. And the same is true for spring floods caused by lots of snow. I conducted an experiment recently—I typed "flood in Russia" in Yandex. Result of my search: "257,858 pages, at least 1157 websites." After that, I typed "flood in France" in the search box. Result: "19 pages, at least 13 websites." And what's amusing, not one reference mentions a word about floods in France! Because there's practically no snow in France today. Accordingly, they don't have spring floods. And what does this combination mean—spring floods, autumnal downpours, and clay soils? O-oh, there's a special Russian term to define this—*rasputitsa*, which means "season of impassable roads." That is, lack of roads is one of the characteristic features of Russia. In the fifth century BC, a messenger of the Persian king Darius moved along the Tsar Road at a speed of 380 km/day (236 miles/day). Ancient Roman messengers also galloped at about the same speed along their wonderful roads. In seventeenth-century Russia, postal service was created with the help of foreign specialists. And the Russian postal couriers moved about the country at a speed of 60-80 km/day (40-50 miles/day). And what is communication? It is civilization. Russia reminded a dinosaur whose nerve signals moved about the huge body very, very slowly.

The situation was saved by the dense network of rivers: wide large rivers in meridional direction and their numerous tributaries in latitudinal direction. Between the river basins were short portages. Up to the latter part of the nineteenth century, majority of commodities was transported in Russia by water—the famous painting *The Volga Barge Haulers* by Repin is a good illustration ... Even after the Romanov dynasty was replaced by the Bolshevik dynasty, comrade Stalin kept with the ancient-old tradition and paid major attention to canals.

Strictly speaking, impassable roads objectively hampered bringing vast territories into a single whole, and therefore the methods of bolting them together needed to be especially strong. But we'll discuss it later.

During a brief and relatively warm period in the early sixteenth century, Muscovites were living mostly in the zone of mixed forests. But as soon as the planet started cooling down in mid-century, Russians began to crawl away like cockroaches.

The slash-and-burn agriculture, which was widely practiced in Russia all the way through the end of the sixteenth century, had quite a low efficiency and required at least one hectare (2.5 acres) per mouth. When population became excessive, Russia began sprawling south and east. Its unstoppable expansion was none other than the spread of the extensive method of conducting agriculture. But what hampered the appearance of the intensive method? Why did Russia not go the Western way?

Starting with the thirteenth century, harvests in Europe were gradually becoming higher. One of the reasons for this (besides the relatively warm and stable climate) was the development of cities. A city is the consumer; it does not produce food, but it does produce civilization. Since cities don't make food, they are ready to pay for food. And as soon as demand appears, supply comes about. Natural conditions in Europe enabled local peasants, using more efforts and brains, to grow excessive grain for sale.

The Harvard professor Richard Pipes, specialist in economic history, once wrote, "Civilization starts only when the sowed grain reproduces itself at least fivefold." By the seventeenth century, European countries reached the "self-seven" harvests and the advanced England managed to grow "self-ten." At the same time in Russia, the average harvest never exceeded "self-three" . . .

What does this mean for the economy? It means that there's no economy. It was not for nothing that one of the European economists said, echoing Pipes, "In a country with a relatively low crop capacity it's impossible to have highly developed industry, commerce, and transportation." And, we can add, following from this it makes impossible to have a superstructure over this economy—the developed political life . . . Let's assume you have land. This is not bad. And let's assume it's in the heart of Russia. This is worse. And let's as well assume that you must farm this land in accordance with the two-hundred-year-old technology. That's the end of you. One third of your land is immediately taken out of production to lie fallow. I understand that the word "fallow" means nothing to a city reader. So I'll try to explain it in a way clear to even a city dweller.

Plants build themselves of those microelements that are in the soil. Russian soils—well, you know what they are. The soil like that—it would be better if it not existed at all . . . Cereal plants suck out everything from this poor soil quite quickly, and after two or three years you're risking of having nothing to harvest. Therefore, the land needs to have a rest, accumulate some microelements. And so this is called "to lie fallow."

So, one third of all the land lies fallow. We get three grains from one sowed. One is left for seeds. And the rest is used for honest starvation and support of a tiny but spiteful government. It's tiny because the soils are poor and unable to feed a greater number of drones. Yet it's spiteful because it's tiny: it needs to bite badly so as the free plowmen would not crush it, the little one. That's why free plowmen are a contra-indication for this government. And they simply could not even survive economically, for there is no economy.

As to "honest starvation"—it might be too strongly said. Let's say it softer—"constant undernourishment." The trouble is not in starving per se, but in lacking any surplus. Russia fell into a vicious circle. Low crops mean lack of surplus that could be used for trade. And economy is primarily trade. Essence of the economy is moving goods along the channels of supply and demand.

Poverty has two dimensions. A poor peasant has nothing to sell, and consequently he has no money to buy something from a city artisan. Accordingly, artisanship is not developing. And since there are no artisans, there are no buyers, meaning there are no incentives to develop agriculture and increase harvest through innovations. These are the reasons why cities in Russia were not centers of freedom (which stems from free trade), but rather centers of military and administration. In the late eighteenth century, when Europe was moving full steam ahead toward citifying itself, the city population in Russia constituted only 3 percent of the total population. And significant part of this city population represented nobility and gentry, that is large or small landowners—the same agrarians but higher in ranking.

The surplus grain accumulated only in the hands of landowners. But they, too, had nowhere to sell it: the peasants could not buy it even if they wanted because of their poverty and lack of money, while Europe had more than enough grain of their own. Russian landowners started mass supplies of grain to Europe only in the nineteenth century—and only because Europe reached that level of its economy when it was cheaper for them to buy Russian grain than produce it themselves. Thus, Russia became Europe's raw-material appendage. However, it had this status in the past, since it was selling to the civilized countries such goods as hemp, mast lumber, badger fat, furs, wild honey . . .—in general, all that grows or produced by nature and not by Russians.

But it should be said in all fairness that Russia attempted several times to break this vicious circle. For example, the circle of the Russian vice could have been broken by expanding foreign trade: let the cities grow on the basis of trade with the external world rather than with the internal peasantry. And when cities would grow, the demand would appear, and the Russian peasantry could then try to meet that demand by way of introducing innovations into agriculture.

The first initial rapid development of cities in Russia happened in the ninth and tenth centuries. That was when Russia got its nickname Gardariki (country of cities). The reason for the growth of Russian cities in that time is simple. Do you remember what the Arabic historian Ibn Khaldun wrote? "Christians could not lower on water even a simple board." Muslims became then the masters of the Mediterranean, and thus the way from the North Europe to the Middle East lay through Russia and its water arteries. However, the "life in clover" soon ended: during the next warming maximum, the steppe dried up and the Turkic nomads, squeezed out from their habitat like toothpaste from the tube, cut this trump trade route with their invasion.

The second upsurge of city life in Russia happened between thirteenth and fifteenth centuries, when Novgorod was the member of the Hanseatic League. It was truly a great chance for Russia. Novgorod was then a super-city with a population of four hundred thousand! As a note, by late nineteenth century, Novgorod had no more than twenty-one thousand. That's what trade means! Alas, at the end of the horrible fifteenth century, this beautiful sprout of civilization on the ugly body of backward Russia was trampled by Ivan III, "the Great," with unprecedented cruelty. Nevertheless, this act up to now is unambiguously treated as positive in strengthening Russian centralized state . . .

The third attempt of revival began in the mid-sixteenth century, when the English came to awaken Russia with trade by way of opening a trade route not through Danish straits of the Baltic but by northern seas.

Cities began growing rapidly then in Russia along the rivers and roads connecting the Russian capital with the Arctic Ocean. However, by the seventeenth century, this meager trade died. As wise people wrote, this happened "partially because the Russian government under pressure by own merchants revoked privileges it had granted to foreign traders, and partially due to the drop of demand for Russian goods."

So what happened to the newborn Russian cities? They fell into decay, becoming places of abode for the tsarist bureaucracy, armies quartering, and entertainment for local landowners. In general, cities founded by the Russians were none other than stockade or fortress towns—to ward off attacks of steppe people. For instance, the entire southern steppe occupied and cleared of aboriginals was enclosed with stockades.

So we ask once again—what prevented Russia from going the civilized Western way?

Vast spaces. And the cold.

The principle of minimum activity on the social level works as follows: no one is going to develop if it's possible not to develop and just live as in old days. If it's possible to expand further and further east while burning down the forests to make place for land, no one will "burn candles" to invent something new. Russians never "burned candles," but just moved east. As to Europeans, they had nowhere to move: population density in Europe was up to its limit then (considering technologies of that time). Therefore, Europeans had to get their heads involved. For instance, invent the plow.

In Russia up to the latter part of the nineteenth century, the main tool of a plowman was a primitive wooden plow that was used probably back in the Neolithic times. It was by no means a European plow that turned over the soil layer, but rather a stupid curved thing that scratched the earth no more than 10 cm (4") deep. But this wooden plow did not require much strength to pull it (the animals were barely alive) and it allowed tilling the field ten times faster than with a real plow (they had to rush: "one day feeds the whole year"). It was not for nothing that the German scientist Storch, who visited Russia right after the Napoleonic wars noted that in no European country, "agriculture is done so remissly."

The main crop of Russian peasants was rye as the most undemanding cereal. But as to the fact that of all various grains rye produces the lowest yields . . . well, God giveth and God taketh away. Similarly, Russians were not eager to quit the three-field system—the inefficient system abandoned back in the late Middle Ages by countries with developed agriculture.

Here's a small example of the "economy" of Russian agriculture. In the eighteenth century, the total amount of work for cultivating one dessiatina (2.7 acres) cost 7.6 rubles in cash. That was the market value of workforce. And the market cost of the goods produced on that same area even at a fabulous crop of "self-six" was exactly a half of that! And if the crop is average—"self-three," the prime cost of the crop becomes four times higher than its market cost! In other words, agriculture in Russia is economically unprofitable and can use slave labor only.

That's why the land was not regarded as a means for enrichment even by landowners: low crops, limited market. Not a single great estate of the Russian wealthy had its roots in farming. As to the peasants, they simply hated the land and the land work and dreamed to go away to some city to do something, become a peddler, artisan, barge hauler . . . anything, but working on the land, for this was hell.

In the mid-nineteenth century, a German agronomist, Haxthausen, visited Russia. He made some comparative calculations of revenues from two farms of the same area in Russia and Germany. The following conclusion is the result of his analysis: if you were presented with a piece of land in Russia for farming, it would be best for you to decline this extremely ruining present. In the late nineteenth century, another specialist in agriculture, a Russian, Engelhardt this time, came to the same conclusion, by saying that the capital invested in state bonds brings greater revenues than the same money dumped into agriculture.

Since agriculture in Central Russia is unprofitable and could be run only with the use of slaves, all estates of landowners in Russia turned into "vertically-integrated companies." They used the slaves to produce not just grain but also various handicrafts like fabric, tableware, musical instruments, harness, furniture (artisans were practically nonexistent). In essence, they represented huge natural economies that were totally excluded from the economy (monetary turnover) of the country.

Here's an interesting illustration to the thesis of agriculture inefficiency and grain market limitations in Russia. For instance, what did the landowners do with the grain surplus? They made vodka from it. What they were actually doing was true art, since it couldn't be repeated industrially due to the same unprofitability. Here's a recipe of typical landowner vodka: to make a ton-plus (or rather 1,200 liters or 320 gallons) of home-brew you would need 340 kg (750 lb.) of grain, some water, and a 12-liter (3-gallon) bucket of brewer's yeast. Plus, a bucket of milk to refine the prime fraction, a couple of young birches reduced to charcoal (for filtration), potash (potassium carbonate), and a couple of cubic meters (70 cubic feet) of firewood for distillation. These 1,200 liters (320 gallons) of home-brew yield only 15 liters (4 gallons) of refined 99.8 percent-pure alcohol.

If landowners were to sell this marvel, made of grain bought at the market price, no one would ever buy it because of its incredibly high and totally nonmarket prime cost. It's not surprising—the yield of the final product is just over 1 percent! But since all of the components—grain, serf labor, lumber, and milk cost landowner nothing, all of noble Russia produced nice vodka, and each estate had its own unique brand and especial refining recipe. Vodka was of such high quality that Catherine II, "the Great," was not ashamed to send it to both Friedrich the Great and the King of Sweden, as well as to her intellectual friends—Voltaire and Goethe. Biologist Carl Linnaeus, having been treated to Russian vodka, became so enthusiastic about it that after having a good sleep wrote a whole treatise dedicated to Russian vodka and the benefits it provides to human body. I'm envious.

So in summary, starting with the fifteenth century, Russia began intensive capturing of foreign territories, mainly because Russia needed land. Just land per se.

Chapter 3

Very Vast but Nonsensical . . .

In the seventeenth to eighteenth centuries, over two million Russian migrants moved from Central Russia to the steppes won from the steppe savages. The largest wave of migrants poured into the Black Earth Belt after Russia occupied the Crimea. It must be said that Russian peasants were not too kind to the local population that was ruthlessly driven away from their lands.

In the nineteenth and early twentieth centuries, thirteen million more Russian migrants left overpopulated Central Russia and moved south, while five million more migrated to Siberia and Central Asia.

The spread of Russian peasantry resembled a triangular flag extending eastward in a form of a narrowing tongue. Soon the narrow tongue of this flag extended along southern Siberia all the way to the Pacific Ocean. And as the Russians were advancing south and east, their communal psychology experienced true miracles. The warmer and mellower was the climate in places where peasants settled, the more individualistic were the motives in their actions, ideas, and behavior, and the less collectivism and communality remained in their minds. While poor soils of the north were cultivated by joint efforts of the entire community, the farms in South Siberia and Black Earth Belt were more of a private individual nature.

As it turned out, the vaunted collectivism and communality of a Russian person were not an immanent feature but rather a climate derivative. Thus, it's not inherent in Russians one and all, but only in those involved in a certain field of activities and living in a certain geographical environment. That field of activities is farming, and the geographical environment is Central and North Russia.

. . . Today, fortunately, there are fewer and fewer Russians of this kind in Russia. They are disappearing. Firstly, because agriculture in these areas is unprofitable under conditions of economics. And secondly, there is no need to have many peasants in an urbanized country . . .

By the way, extensive swelling of the Russian Empire wasn't at all heroic-pioneer in nature, as it's usually depicted in school textbooks. Russia's expansion was driven by thugs who would never be welcomed by a present-day urban citizen. For instance, in the city of Khabarovsk (in the Russian Far East), people like to lay flowers at the monument to the pioneer Khabarov. And practically no one knows that this pioneer and the members of his expedition, while sailing along the Amur River, hunted heatedly along the river shores for local population that was used as food. Simply put, they were cannibals. It was easier to hunt people than game.

Other pioneers were not much different from Khabarov. For example, ancient citizens of Novgorod, who successfully combined trade with robberies, mercilessly abused and knifed northern savage population. Like all gangsters, they knifed them under a noble pretext—to baptize Lapps, Karelians, and "bloody self-eaters."

After Novgorod was defeated by its "competing organization"—Moscow, the capital undertook upon itself the noble mission of robberies. Muscovites approached this business monumentally—they began establishing stockades and monasteries in tundra. The founder of the world's most northern monastery in Pechenga was some Father St. Tryphon. The diary of a Dutch merchant, Simon Van Salingen left a brief characteristic of St. Tryphon for future generations. Before becoming the commander of monks and live personification of Russian presence in this wild tundra, Tryphon, according to his own cynical admission, "robbed and ruined many people and shed lots of blood." Besides that, the Father liked to put his lips to a glass. Or rather a mug . . .

His monastery was involved in quite a God-pleasing business—trading fish and salt and exploiting the simple-minded Lapps (a.k.a. Sami) in the cruelest manner. Aboriginals were exploited in the name of God so mercilessly that the researchers studying the Lapps life noted that the Pechenga Monastery "was for Lapps an utter calamity." Well, what else could be expected of former robbers and murderers who reached such a state in their dissipation that a special church commission from the capital came to investigate their actions?

The protocols of the commission honestly stated, "The monk Ilya lives his life totally drunk and steals monastery goods wherever possible, and he took monastic vows only to escape troubles caused by his thefts." And there were others like Ilya. Here are some characteristics of other Russian civilizers: "he lives his life totally drunk, barely leaving the tavern," "he's a booze-loving person," "likes the booze in large quantities," etc. Apotheosis of the church commission investigation came in the following verdict: no longer "would women be invited or allowed in the monk cells for any purpose"!

Resulting from all its conquests, Russia had so much land that quite unexpectedly for itself it took the first place in the world with respect to the size of its territory. However, the perpetual Russian fate named Cold played an evil joke on Russia. The major part of its territory is simply unfit for habitation. Two thirds of Russian territory is permafrost. And it was observed long ago that civilized people acclimate poorly at altitudes over 2,000 meters (6,500 feet) above sea level and in areas north of annual average temperature -2°C (28°F). It's unbearable for normal people to live in such wretched places! That's why mastering of the Arctic by white people happened only in periods of global warming when at least it was possible to stick the nose there.

Both the famous Pustozyorsk stockade in the Pechora mouth and the mentioned Pechenga monastery were founded in periods of brief Arctic warming. In the sixteenth century, Russian people eagerly flocked north. It was believed previously that people were going north under repressions from their terrible tsar nicknamed the Terrible. Perhaps. But why run away from the Terrible into Lapland? The reason was that at this time—from 1530s through early 1570s—a global warming happened, making life possible at those latitudes. The average decade temperatures in this period at northern latitudes were 2°C (4°F) higher than temperatures of the coldest decades of the fifteenth and late sixteenth centuries. And this is equivalent to moving 600 kilometers (370 miles) south. In other words, climate in Lapland was at that time the same as 240 kilometers (150 miles) north of Moscow.

But in the end of the sixteenth century, when the temperature dropped sharply and the north again became north, the Russians were no longer in a hurry to settle not just in Lapland but also in the more southern

White Sea area. Moscow even had to adopt repressive measures to make people settle in the basin of the Northern Dvina, where on orders of the progressive tsar Boris Godunov the city of Arkhangelsk (Archangel) was founded with the purpose of trade development.

In summary, in its hunt for land Russia grabbed so much of it that it became the largest country in the world. One sixth of the global land mass! At the same time, life was impossible on 70 percent of this territory. Well, it's not for nothing geographers have a term "effective territory," or the area of a country where life is possible. Full happy life of a civilized person, but not a tribal savage hunting seal with a spear. Using this criterion, Russia is far from being the largest country in the world. Its largeness turned out hollow.

In this respect, it would be interesting to look at the following table 2 showing some countries' territories both geographical and effective.

Table 2

Geographical and Effective Territories of the World's Ten Largest Countries in 1900 and 2000 in million sq. km (million sq. miles)

Territory in 1900	Territory in 2000	
Geographical	Geographical	Effective
1. Russia 22.47 (8.68)	1. Russia 17.08 (6.59)	1. Brazil 8.05 (3.11)
2. China 11.41 (4.41)	2. Canada 9.98 (3.85)	2. USA 8.00 (3.09)
3. Canada 9.98 (3.85)	3. China 9.60 (3.71)	3. Australia 7.68 (2.96)
4. USA 9.66 (3.73)	4. USA 9.36 (3.61)	4. China 5.95 (2.30)
5. Brazil 8.36 (3.23)	5. Brazil 8 .51 (3.29)	5. Russia 5.51 (2.13)
6. Australia 7.69(2.97)	6. Australia 7.69(2.97)	6. Canada 3.64 (1.41)

7. British India 4.83 (1.86)	7. India 3.29 (1.27)	7. India 2.90 (1.12)
8. French W. Africa 4.69 (1.81)	8. Argentina 2.78 (1.07)	8. Kazakhstan 2.62 (1.01)
9. Osman Empire 3.90 (1.51)	9. Kazakhstan 2.72 (1.05)	9. Sudan 2.49 (0.96)
10. Argentina 2.78 (1.07)	10. Sudan 2.51 (0.97)	10. Argentina 2.45 (0.95)

It's evident that Russia isn't the largest country at all, but only the fifth in area when the "territorial slag" is discarded. And it lost half of its effective territory in the twentieth century when its southern and western, or warmer, regions fell off.

The talks of Russia as the world's coldest nation have palled on all to the extent that seem commonplace. Many probably know the book by Andrei Parshev, *Why Russia Is Not America*, which discusses in detail the climate peculiarities of Russia. But many citizens still have some strange misunderstanding here.

The difference in climate between Russia and the West, which led to differences in mentalities, religions, ways of life, etc., is so striking that it's impossible not to draw attention to it. It's in temperature, moisture, insolation . . . If you were to find somewhere a map of radiation balance, you would see with your own eyes that the Lord deprived Russia not only of temperature and precipitation but also of light. There's as "much" sun in London and Paris as in sultry Tashkent—60 to 80 calories/cm² (390-520 calories/in.²). And the isorad with a number 40 (260 calories/in.²) is the one separating forest and steppe zones of Russia. We live in "darkness." And only having subdued the steppe inhabitants, the Russians managed to cut their way to sun light.

The media published more than once a table of annual average temperatures of the world's coldest countries. Russia leads in this table with a big margin. I think it's no sin adducing it here once again (table 3).

Table 3

Annual Average Temperatures of the World's Coldest Countries

Country	Annual Average Temperature, °C (°F)
Russia	-5,5°C (+22.1°F)
Canada	-5,1°C (+22.8°F)
Iceland	+0,9°C (+33.6°F)
Finland	+1,5°C (+34.7°F)

However, even seeing these figures, many people are still unconcerned: "But these are average values! The negative temperatures come from cold places like Siberia where almost no one lives. And also, in Canada the temperatures are almost the same and forty-degree frosts (-40°C=-40°F) are not infrequent there."

Oh well, let it be Canada. Indeed, Canadian north does experience frosts even worse than minus forty. But in those areas, where frosts reach minus forty, nobody lives except some rare persons on special duties. People prefer living in cities. Toronto and the Canadian capital Ottawa that's somewhat north of it are located on the same latitude as Russian Black Sea cities Tuapse and Anapa. And the most populous Canadian northern city Edmonton is on the same latitude as the Russian city Oryol that is 300 kilometers (190 miles) south of Moscow. I'll repeat: this is the northernmost city with one hundred thousand people! Three times less than in Oryol.

But civilized people just don't live up north! And only Russia is the sole country in the world, where in the zone not meant for life (like north of the notorious isotherm -2°C [28°F]), cities exist with populations exceeding one hundred thousand—Surgut, Vorkuta, Inta, Nizhnevartovsk, Norilsk . . . And nearly all of them were born in the era of the empire. It's not surprising that today, when the empery weakened, former civilizers began moving south from north and east.

Russian north loses its population rapidly (table 4). Outflow of people from regions with the most extreme natural conditions took a truly plummeting rate: for example, Magadan province lost over a half of population, but the absolute record belongs to Chukchi area whose population decreased to one third in just twelve years!

Table 4

Population of Some Districts and Subjects of RF in 1990 and 2002 (in thousands)

Federal Districts and Federation Subjects	1990	2002
Far East Federal District	8,017	6,693
Amur Province	1,059	903
Jewish Autonomous Province	216	191
Kamchatka Province	476	359
Koryak Autonomous District	40	25
Magadan Province	390	183
Primorsky (Maritime) Region	2,279	2,071
Sakha (Yakutia) Republic	1,112	949
Sakhalin Province	714	547
Khabarovsk Region	1,611	1,436
Chukchi Autonomous District	162	54
Siberia Federal District	21,113	20,063
Agin Buryat Autonomous District	77	72
Altay Region	2,641	2,607
Irkutsk Province	2,792	2,582
Kemerovo Province	3,096	2,899
Krasnoyarsk Region	3,161	2,966
Novosibirsk Province	2,741	2,692
Omsk Province	2,153	2,079
Buryat Republic	1,045	981
Tyva Republic	313	306
Khakassia Republic	571	546
Taymyr (Dolgan Nenets) Autonomous District	55	40
Tomsk Province	1,084	1,046
Ust-Orda Buryat Autonomous District	137	135
Chita Province	1,320	1,156
Evenk Autonomous District	25	18
Urals Federal District	12,737	12,374
Kurgan Province	1,107	1,019
Sverdlovsk Province	4,774	4,486
Chelyabinsk Province	3,705	3,604

Today, the-once huge empire shrank, with respect to its population and effective area, to the dimensions it had when it started on its great path in the sixteenth century, holding proudly and high the willfully painted Roman flag ("Moscow is the third Rome, and there will never be the fourth!").

Will the collapse of the empire stop here?

PART 7

Collapse of the Empire

With stars, ceramics, looks, and prayers—
With all these means to see the trend,
How was it possible, soothsayers,
For you to not foresee the end? . . .

Igor Tsaryov

"From all we know today, is it possible to make a conclusion that in eras of global warming the collapse of great nations is inevitable?

"Well, that was always the case," Klimenko shrugs his shoulders. "Look at the climate fluctuations graph superimposed over Russia history. As soon as climate cooling or drying up occurs, we immediately see trends of unification and rallying of the country and the nation, spiritual uplift, and capture of new territories. All periods of maximum territorial expansion of Russia and its consolidation coincided with the eras of climate worsening. But just as the warming arrives, centrifugal trends immediately come into action and Russia disintegrates like a house of cards. Of the twenty-eight historic events that could be attributed to strengthening of the state, twenty-one events took place in periods of cooling and three in periods of intensive drying up that, too, were climate-worsening periods. In other words, it's the climate, which controls ambitions of power.

"Take, for example, the twelfth century when Kievan Rus collapsed after the death of Mstislav the Great. The twelfth century was the period of the most propitious climate in the territories of modern-day Russia. And by the way, it was very much like the one we're having today. After that, the climate worsened sharply, and Russia was consolidating—under the rule of Ivan I Kalita; Dmitry Donskoy; Vasily II, "the Blind"; and Ivan III, "the Great." During the reign of Vasily III, the father of Ivan IV, "the Terrible," the warming came, and Russia immediately lost some of its European territories to Poland, Lithuania, and Sweden . . .

"I have an interesting quote here—right to the point. In his address to the Federal Assembly in 2005, Putin said, 'First of all, it's worth noting that the collapse of the Soviet Union was the major geopolitical catastrophe of the century . . . At the same time, the epidemic of collapse spread into Russia itself . . . You know that in the last five years we had to solve tough problems on preventing the degradation of state and public institutions.'

"I'm afraid all his efforts directed against the climate would be wasted. It's impossible to oppose sunrise or sunset, or the change of seasons. Humans are weak . . ."

CHAPTER 1

Colossi on Clay Legs

The latest global warming, which is the talk of the day, started approximately at the time when humans began to actively return into the atmosphere the carbon that had been once taken by the flora and deposited in coal and oil. The industrial era started, humans began actively burning fossil fuels and building factories, thus emitting into the air the carbon in the form of carbon dioxide. The global temperature was actively moving upward since the 1920s when the iron steed was gradually replacing the fetid peasant horse.

All the empires, existing at that time on the planet, began cracking like the Perito Moreno Glacier and breaking down to pieces. While there were fifty-two independent states at the start of the twentieth century, by its mid that number rose to eighty-two and today, it exceeds two hundred! The last one to collapse was, naturally, the northernmost and coldest empire whose name does not even need to be pronounced—every reader knows it well.

Empires are no more. There are countries, there are their former colonies, there are influence zones of big countries, but empires per se or colonies are no more. But there still are big countries. Big in the territorial sense. Won't these big ice floes break up due to global warming and its inherent centrifugal effects?

If a system is rigid, everyone walks in formation and obeys orders from one center. A system like that is good to withstand external enemies and forces of nature, but it's not flexible. And a flexible, adaptive system can't be rigid. Its structure is much more complex. What "more complex" means? It means it has more centers that make decisions.

An iron rod is an iron rod. A thing consisting of one element. And a flexible blade of grass consists of a myriad of live cells with their nuclei and organelles . . . As much as a blade of grass is more complex than an iron rod, so much is the modern postindustrial society more complex than the primeval empire. It's true that an iron rod could break someone's head, while the blade of grass is very weak indeed. But it has more life in it. In full compliance with synergy principles, complexity of social organisms increased along with increase of energy they consumed. In the last two hundred years, civilization increased its energy consumption fivefold. This led to the twofold increase of life expectancy, to the reduction of workweek by half, and enabled to provide food products for the planet's sixfold-increased population. That's what energy means! In the contemporary world, it's the per capita energy consumption, which is the factor determining the country's status and its development level.

It could be heard occasionally, "Well, we have plenty of energy in Russia! The country is full of natural gas, oil, and various minerals!" That's true. But we remember how the Russian bragging about its vast territories ended. Two thirds of them turned out to be a useless acquisition. Could the same happen to the entrails of our land? It could possibly turn out that it would be simply unprofitable to extract or transport some of the deposits because of their overly cold and/or overly remote location. Under conditions in Russia, even gold could become unprofitable . . . As to the deposits or energy resources, such as oil, natural gas, and coal, it would be worth looking at how much of them is inevitably spent on simple survival under conditions of the terrible climate and how much could possibly be spent usefully—like for manufacturing or selling.

The world's richest countries (USA, Canada, Norway) use per capita 10-14 tons of equivalent fuel a year, while the poorest countries consume about half a ton. A ton of equivalent fuel is a comparative rated value enabling to bring various fuels (oil, coal) to a common denominator. One ton of equivalent fuel equals 7 billion calories or 29.3 billion joules. That much energy is produced by burning one ton of high-quality coal.

But energy consumption is not yet the sign of richness, nor is double window frames in Russian houses. It's in the warm England, where double window frames are the sign of luxury that is mentioned when selling real estate. And they are never mentioned in Russia because there's nothing else. And it's not for the reason of being rich but rather out of necessity, for it's hard to survive otherwise.

If we were to compare Bulgaria and Spain on the basis of per capita fuel consumption only, we should consider Bulgaria a higher developed country because it consumes 15 percent more of equivalent fuel than Spain. And the poverty-stricken Romania is even more advanced, consuming 20 percent more than Spain! But in fact, Bulgaria lags behind Spain in GDP by one order of magnitude, while Romania lags even more.

Why is it? The point is that these two countries are east of Spain by far. Or closer to Russia. Meaning closer to the Cold Pole (which is, as you remember, not at the North Pole but in Russia). Thus, both Bulgaria and Romania must heat themselves. And if the fate forces Bulgaria, a southern resort from the Russian perspective, to use valuable fuel for heating, then what's to be said about the world's coldest country?

Well, here's what. Half (!) of all extracted "fossil fuel" (coal, oil, natural gas) has to be simply burned in furnaces—that is, used for elementary physical survival under conditions of cold climate. All this energy doesn't add one iota to quality of life, compared to southern countries, but just makes living conditions bearable indoors where Russians spend most of their lives. Even rather than not adding anything, it actually takes away, because we have to pay for heating. And we have to pay a lot—look at a monthly apartment payment bill. We have to pay in order not to croak. Some climate racket!

As it turns out, some can have fuel under their feet and use it for heating, or have none but use sun for heating. In other words, northerners compared to southerners must always spend more energy and money to ensure the same quality of life. Here's an example. Iceland and Malta have a comparable quality level of life. But the annual average temperature in Iceland is 0.9°C (33.6°F), whereas in Malta it's 18.5°C (65.3°F). So Iceland uses 9 tons of equivalent fuel per capita per year, while Malta uses only 2.5 tons. Sense the difference! . . . Icelanders are fortunate to have "free" energy—they have hot geysers under their feet, which can be used for heating and for greenhouses, although substantial efforts and resources are required. Maltese are fortunate, too,—they have lots of sunshine and so greenhouses are not needed at all. Warm climate is an additional economic resource of the state, same as oil or natural gas. But abundance of coal and other minerals in Russian entrails is unfavorably compensated with its amazing climatic poverty: as we remember, Russia is the world's coldest country. And it's not the only minus of Russia.

The other minus is exactly the fact that has been used as the source of pride—the country's vastness. If you have ever heard about the law of big state inefficacy, you might have guessed the next topic. Indeed, the big state is non-optimal from the energy point of view. And it should be discussed in detail . . .

Everyone is bored stiff from the phrases that the world is in transition from industrial to postindustrial phase of development. But few understand what it means. Yet some citizens with bits of information respond quite competently that the postindustrial era differs from the previous one in that civilization shifts its emphasis from industry to information technologies, entertainment, and services. Good answer. An "A."

But from the energy point of view it means that postindustrial societies gradually stabilize consumption of energy per each individual (fig. 8).

The graph shows that after the energy crisis of 1970s, energy consumption dropped sharply (it's especially pronounced with respect to the largest consumers, like Canada or USA), and the world began the transition to energy-saving technologies. After that, the rise in per capita energy consumption slowed down, showing a trend toward stabilization. It's clear: a person doesn't need three juicers nor can simultaneously drive two cars or excessively use the furnace in winter by raising the temperature above comfort level. The majority of the population in postindustrial countries represents the middle class. And a "middle-class-man" won't buy a hundred-room house that would be impossible to maintain. This person would not take a two-hour shower wasting water and time, since water costs money and time is needed to work. At the same time, all basic needs of life and entertainment of a postindustrial person are satisfied. It's hard to think of something power-intensive. DVD replaces VHS, one juicer replaces another, but all of this in no way increases energy consumption in the house. On the contrary, the household equipment and cars become more economical. However, this applies to "proper" countries. As to "improper" countries, where the main entertainment is to shoot from AK, it's too early yet to talk not just of their being postindustrial but just plain civilized in general.

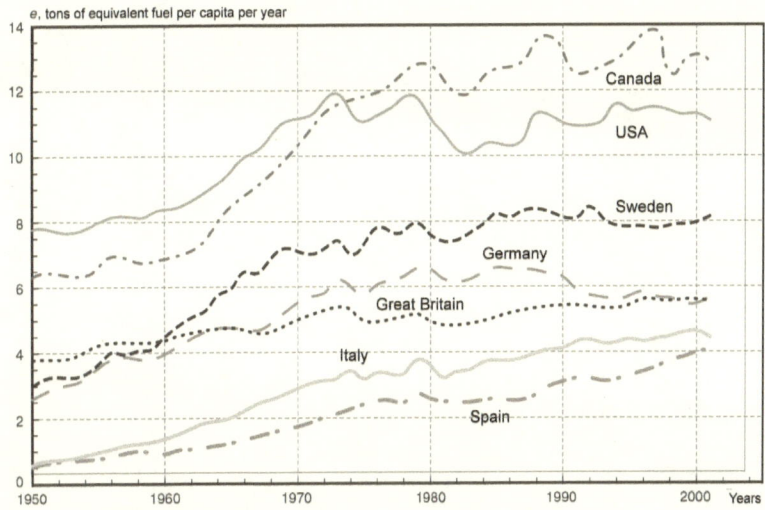

Fig. 8. Energy consumption per capita in some
developed countries

Speaking of civilization—today, in relation to Islamic extremism, it became fashionable to mumble about conflict of civilizations. I have partially written about this mythical conflict in my book *Civilizer's Fate*. And now is a convenient occasion to continue that discussion.

Please pay attention to the graph (fig. 9). It shows dependence of specific energy consumption on the annual average temperature. The countries shown are very diverse—with respect to both culture and economic development. But nearly all of them fall within a 20 percent margin on the mean line that breaks at approximately 17°C (63°F). It means that regardless of cultural differences all social organisms conform to general laws.

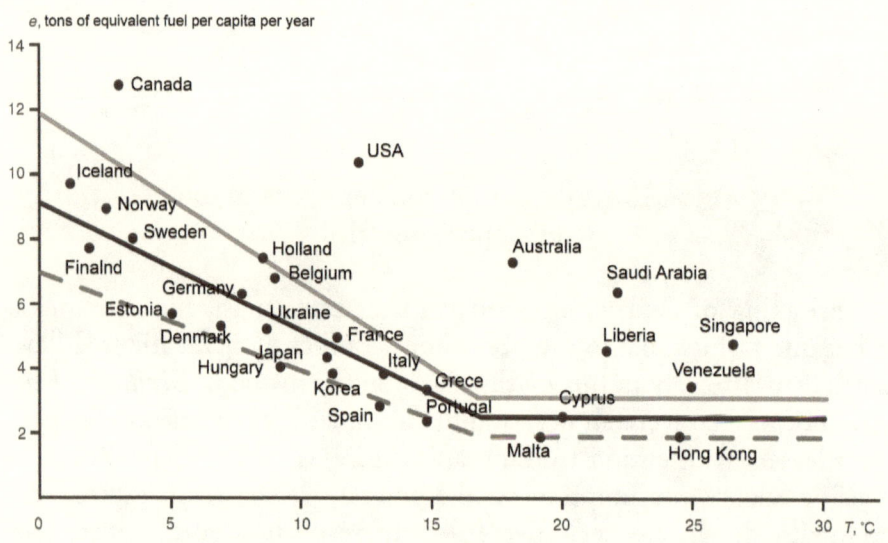

Fig. 9. Correspondence of specific energy consumption
to the average annual air temperature

What does this graph tell us? Firstly, that the energy consumption drops with the rise of temperature, for the need in heating declines. In places with annual average temperature approximately equal to indoor comfort temperature, heating is not needed at all. And secondly, the graph shows that several countries do not conform to general laws. These are large countries with area at least 0.9 million km² (0.35 million sq. miles). Let's note it.

An exception is tiny Singapore, which for reasons unknown got in line with large countries. But this could be explained. The point is that officially the population of Singapore is about three million. But every year about thirteen million foreigners visit this city-state. Besides, Singapore is the world's largest port providing bunkering, or filling the ships with fuel. So if we divide total energy consumption by the total number of not official but real people, Singapore will also get within the theoretical range.

It would be even more interesting to show the same graph in other coordinates—energy vs. territory, thus showing explicitly the dependence of specific energy consumption on the country's area. For purity, we'll use not the formal geographical territory but only useful or effective one.

The following graph (fig. 10) has the effective territory laid off along horizontal axis and the ratio of actual specific energy consumption to the optimal one along the vertical axis. The graph shows that starting with a certain size of a country its energy consumption begins to grow. It grows regardless of the country's economy, political system, culture, as well as temperature and relief, because these factors were cleared away from the dependence. The graph shows only the effect of the country's area on its energy efficiency.

Conclusions? They are the same. Firstly, all postindustrial countries fall under the same system of equations, meaning that they follow the same laws. Secondly, if the country's effective area exceeds half million square kilometers (0.2 million square miles), the country starts using its resources inefficiently (that area is about the area of Spain and France combined).

In other words, if a country is big—it's inefficient. And energy-inefficient systems must collapse. That's the physics of our world.

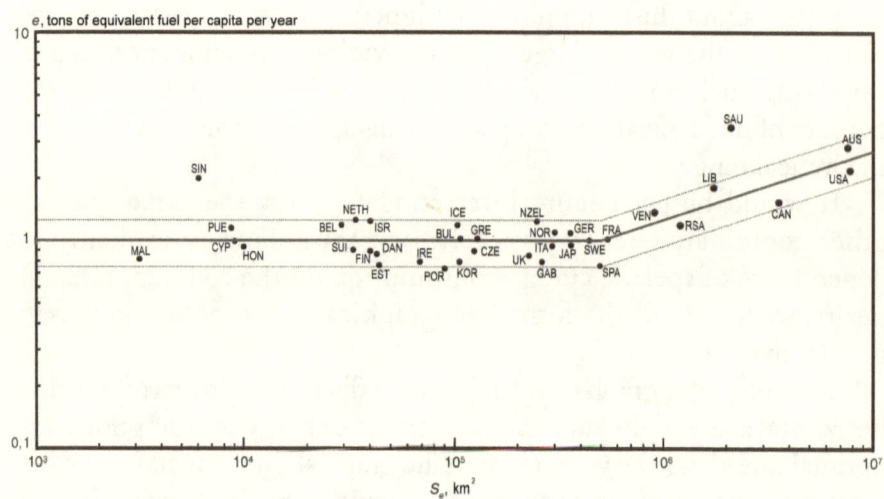

Fig. 10. Correspondence of specific energy consumption
in postindustrial society to country's effective area
(for Singapore, the area is increased tenfold)

CHAPTER 2

Dinosaurs Must Die

Here's a fact: Annual average temperature in the United States and Japan is the same at 11.2°C (52.2°F). Both countries are postindustrial. But per capita annual consumption of equivalent fuel is 11 tons in United States and 4.5 in Japan. 11-4.5 = 6.5. That's how much an American spends to support the grandeur of the country that's too big.

Another fact. The United States spends 50 percent more energy than Europe for manufacture of one unit of goods. The reason is the same.

But why, with all the other conditions equal, a big country wastes more energy than a small one? It's especially pitiful because the wasted energy reduces quality of life for the population that is forced to pay own money for the "hollow grandeur of the country."

Well, firstly, transportation expenses. Let's take Siberia. You can travel by train days in a row along the limitless taiga (boreal forest) vastness and become shocked—oh, Lord, is it really all ours? Alas, it's indeed all ours, and heavy trainloads of coal and other goods have to be transported across it. Hundreds and thousands of kilometers (miles) of tracks and unpopulated space! And recall the scenes in old Hollywood movies of famous American roads extending like endless threads through yellow deserts and prairies with cacti . . .

Secondly, in a big country, some regions are always richer and some are poorer (in Russia this takes extreme forms, with a few donor regions that could be counted on one's fingers and a whole horde of sponger regions). The state, attempting to ensure approximately the same level of life for all, is forced to redistribute the wealth from rich regions to the marginal ones, thus discouraging them both from working. This is also the cost of the territorial grandeur.

Thirdly, people are made what they are by what they eat. And they eat primarily what grows or lives in a given region. Diverse living conditions form diverse viewpoints on life, diverse traditions, behavioral stereotypes, mentality, and accordingly, laws stemming from traditions. Therefore, attempts to rule diverse natural regions from a single center using the same paradigms are doomed to be inefficient from the start. Overcoming this inefficiency turns in the end to losses in energy and quality of life, although the latter argument loses its value more and more. Information exchange (Internet, television), globalization, civilization (spreading the megalopolis principles and the way of life onto the rest of the world) gradually erase nationalities and to some extent cleanse the humanity of regional cultural kinks and inadequate reactions. Yet, it's still hard to imagine today that one and a half billion Chinese would be using huge American cars.

It's known that the postindustrial society, besides stabilizing specific energy consumption, has another characteristic feature enabling to unmistakably identify the developed (urbanized) society: falling birthrate. Is it possible to introduce this criterion into graphs and diagrams? Why not? Let's try . . .

One of the most illustrative demographic indices being discussed in recent years by anyone who would take the trouble is the difference between the rates of birth and death. In Russia, for instance, it's such that Russian population decreases every year by seven hundred thousand people.

The following diagram (fig. 11) shows the world's countries with 99.7 percent of the planet's population (data as of 1990). What's good about this diagram? Well, it represents a simple two-factor space, and looking at it allows immediate understanding of the state of modern civilization. The arrows show the direction where countries are moving. All countries move at different speeds to a single gathering point marked by 1 on the vertical axis, that is, move in the direction of zero population growth and energy consumption optimization.

(The diagram is actually incomplete. It has in addition a left part of negative population growth. Today the developed countries replenish and stabilize their population not through birthrate, but rather through external borrowing—migrants from the third world. But that's a topic for a different discussion.)

For orientation in the diagram's space, all countries can be divided into five main groups (these countries and areas of their habitat are marked on the diagram with Arabic numerals). The closest to the point of gathering are, naturally, the developed postindustrial countries. They have high revenues, low birthrate, and optimal specific energy consumption (their actual energy consumption practically equals the optimal for their climate-geographical conditions). The habitat area of these countries is marked by number 1.

Number 2 shows the countries with transitional economy, such as Russia, for instance. Russia behaves demographically like an adult country—its population propagates less than dies. But the quality of life is still low and the energy consumption is lower than in developed countries. It's obvious that these countries represent mainly former Soviet republics and countries of East Europe—the countries of socialist camp were urbanized enough to stop the population from propagating itself, yet not quite developed economically to have the same quality of life as decent people.

Number "3," the "young tigers," are countries, which are quite developed economically and in the latter decades of the twentieth century made an economic spurt, but still have peasant mentality and thus retain quite high birthrate. They did not yet reach the energy consumption level of developed nations, and therefore their energy ratio is less than 1 and is in the range of 0.46-0.54 (the vertical axis has logarithmic scale).

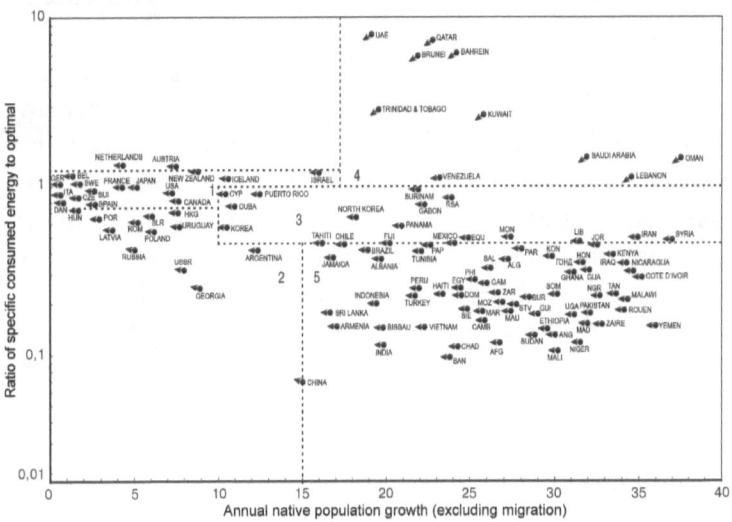

Fig. 11. Energy-demographical diagram of the world

In the upper right quadrant (number 4) are countries that are too well off and fussy. These countries don't earn money. Dollars are simply falling upon them from the heaven, or rather sucked out from the entrails. These countries, as you've already understood, are oil exporters. And the money is sucked out from the entrails not by owners of the land but by alien engineering specialists from America and Europe, doing all the intellectual work. And all the dirty work is done on behalf of these countries' citizens by Gastarbeiter. For lack of anything better, the hosts are left only with buying yachts and long white limousines and feeding terrorism with comp money. In spite of being filled with money up to the ears, customs and tastes in these countries remain totally wild, bumpkin-like, medieval, lacking any faceting by urban high-tech civilization. These are most dangerous countries! Their people with primitive consciousness received free unlimited access to resources, for which they need not to work.

The group number 5 is in the lower right quadrant. Wild or developing countries. Their people propagate in a bumpkin way and live lousy (low energy consumption, far below the postindustrial level).

As soon as all countries (and this will happen sooner or later in accordance with the process theory) will be at the gathering point (I'm tempted to call it the "collapse point," but I'll resist), the phase transition on the planet will be over and the world will enter some principally new stage of history. (For more details on singularity of history, see my book *Russian X-Files*.)

Why was the reader shown all these graphs? So the reader would understand that the point is not that the world is experiencing today conflicts of civilizations—spiritual vs. consumerist, East vs. West, Atlanticism vs. continentalism, etc. Let's leave these tales for Mr. Dugin, a Eurasian. But you can see that the arrows on the diagram are not in conflict, they are all pointed in one direction. It's just that some countries have already reached the gathering point (consumerist society), while the others have not. And that's the only "civilization difference." A little more patience is needed, and Islamic terrorism will disappear by itself—as Islamic countries would be gathering in singularity. And at this point, all multicolored threads of cultures will be interlaced into one plain-looking global information-power cable.

CHAPTER 3

Sadly, the Life in That Era
So Beautiful . . .

(The chapter title is a quote from a poem by a popular Russian poet N. Nekrasov; it continues ". . . isn't meant either for you or for me."—Translator's note.)

Would all the countries be able to get through the eye of the postindustrial needle? Would all be able to improve their quality of life to a level of American life in American clover? Or Japanese at least?

Let's think . . .

In absolute figures, Russia consumes not too little energy, surpassing here Germany, France, and Japan. But we know that the lion's share of this energy is spent not to improve people's life but to heat the atmosphere: first in our homes, and then—as the heat escapes through ventilation, windows, and cracks—the environment in its entirety. Alas, the outer space can't be heated with home radiators. In order under conditions of our climate to raise our quality of life up to the level of West Europe, we have to spend per capita twice the amount of energy, about fourteen to fifteen tons of equivalent fuel annually, rather than seven to eight tons, spent today.

Increasing energy consumption twofold is the same as having two more oil-and-gas bearing areas as huge as West Siberia added to all the oil and gas reserves of Russia. Obviously, we don't have them. The problem is all the more impossible under conditions of so-much-talked-about soon-to-be-exhausted oil reserves. However, it doesn't seem so impossible when humans talk and do ever more about mastering thermonuclear energy, although we can't expect total transfer of global economy to thermonuclear tracks in the next one hundred years. On the other hand, there are geological theories promising free hydrogen right from the entrails of the earth (for details, see my book *Monkey Upgrade*), but this hydrogen, alas, is still only in theory.

But the problem of increasing the specific energy consumption could be solved in a different way—by reducing the population by half. Raise quality instead of quantity! It's a nice solution, but we can face a different problem here. Half of the population might simply be not enough to ensure the workability of the country's entire infrastructure. As soon as the population mass becomes less than critical, collapse may follow.

Another way of improving quality of life and country's energy efficiency is the break-up of a unitary state (preferably under control) into several independent entities. For some reason, the out-and-out patriots express uncontrollable animosity against this particular version. They are ready for anything, even for our people to live worse than abroad, but for the sole purpose to have an option to go out on the porch in the morning, slap the suspenders, and happily exhale: "Grand is Russia!"

I don't know what Uncle Freud would have said about this unmotivated desire to live in a huge country, but I suspect that certain citizens might still have in their consciousness some naive-teenage features. For some reason, patriots have grandeur, consistently associated with size rather than with quality of life. The ancient archetypes that work perfectly in the animal world.

What's the danger if a country breaks up under a process with no war, like the peaceful separation of Czechia and Slovakia, for instance? The danger is in our heads. The childish desire to live in the largest (well, let it be even the fifth largest in effective territory) country is too strongly fixed in the peasant brains of too many of our citizens, making this factor hard to ignore.

By the way, the separation can take place so imperceptibly that the citizens might not even notice the separation process. The reason is that two waves, integration and disintegration, could run into each other in Russia. Centrifugal trends lead to the regions grabbing for themselves so much independence in politics and economy that the country would become de facto a confederation, although formally it would still remain a unitary state. So that's one wave. And the second wave is centripetal, the one that integrates, moving exactly in the opposite direction. What this wave is and how can it occur in the era of global warming and the universal striving for atomization and individualization? I'll try to explain . . .

What is progress? It's the ever-greater deviation from the nature, isolation from it. Civilization replaces the natural habitat with the artificial environment—the technosphere. Technosphere is foremost a megalopolis. The artificial environment is built in accordance with its own laws and is a kind of a damper, or a shock absorber between the natural environment and humans. The technosphere dampens the bursts of natural environment. We see today how nature-political countries that developed from agrarian societies are integrating into a Unified Europe. What is the basis for their integration? Naturally, it's technosphere. And what is the core essence of this integration, in other words called globalization? The second name for globalization is standardization.

Standardization advances in technosphere on all fronts just because it can't not advance. Standardization covers nuts and bolts, formats of audio and video cassettes, and network protocols. Standardization covers jurisprudence—international law starts prevailing over national laws. Standardization covers sanitary and hygienic rules. Standardization covers rules of economic interaction—trade and workforce exchange—WTO appears. Standardization covers political institutions and political structure of a society. Standardization covers finances—a single currency appears for a huge region.

It's safer and more convenient to live and trade in a standardized civilization. You can always be sure of not being poisoned by foreign food because the hygienic rules are standardized internationally. You can always be sure that an audiocassette of one manufacturer would fit another manufacturer's tape recorder. And you can always be sure that your interests infringed by a national government can be protected by a supernational court of Strasbourg. Traveling as a tourist or on assignment, you have no pain in your butt when converting currency from one to another at a border crossing. Standardization is the foundation of unification and integration of techno-plates of different countries into one single technosphere.

Integration can easily replace disintegration processes of a collapsing country upon its reaching some low point, just as a circus gymnast on a trapeze timely catches a flying auntie in air. It can happen during or after the process of Russia collapse. Or it can start right today, on the stage of postimperial collapse, through integration of the post-Soviet space into one economic whole.

In any case, some unification processes can already be observed. The number of independent regions in Russia decreases steadily. As recently as 2005, referenda were conducted on unification issues in five regions: Krasnoyarsk region, Perm, Taymyr and Evenkia, Kamchatka and Koryakia. Ust-Orda Buryatia autonomous district was united with Irkutsk province. Awaiting unification are Yamal-Nenets and Khanty-Mansi autonomous districts with Tyumen province. Some movement is noticed in Komi in search for someone with whom to unite. The nonsensical Jewish autonomous republic—one hundred poods (3,600 pounds)!—will join Khabarovsk region. And it's time in general for Adygea, too, to stop being a wart on a healthy body of Krasnodar region.

... Analysts predict that in the coming years the number of Russian regions would come down from 89 to 60. Well, we shall wait and see ...

The historical experience tells us that all nations and ethnic groups, which have in them the potential for independence, must certainly, prior to the stage of supernational global integration, go through the stage of "youth hypersexuality"—obtain independence in a form of national statehood. It's believed that there's no alternative to this experience. But this process might be well overlapped by the coming process of total globalization that erases national features, as well as by the process of human genetic modernization that acts in the same way.

So if the global warming will continue, if in the near future something catastrophically freezing will not happen, requiring some rigid tightening of screws, then the world of unitary states in a not-distant future will turn into a civil-network world. On the one hand, it means collapse; but on the other, it's economic integration.

But can we be certain that the global warming will continue? What's the temperature's future?

PART 8

Temperature's Future Is Civilization's Future

The wind was blowing brutally from one edge of the endless plain to another, and two persons appeared hopelessly small in the very center of the half-tundra—half-woods, which was so monotonous that each step did neither bring closer to nor hold away from anything. The snow lay here and there, pierced by black wet branches of brush, forming isles, piles, pieces . . . Bleak, vicious world. Not a single place to stay warm—snow . . . But the two, with their slow movement being riveted to that area where they were born, never saw anything else, just heard from the elders that it was better here before. And not the cold worried them; they were children of cold rather than of warm.

Sever Gansovsky

How many winters, how many summers it took us to reach this Spring?

Igor Tsaryov

Well, that's how it happens! Just as I'm writing these lines, the temperature in Moscow fell in two days from 0 (32°F) down to -30 (-22°F). Rain changed to vicious blizzard. Meteorologists are saying that the severe freeze would stay for the whole two weeks.

Until now, winter temperature was dancing around zero (32°F). And the fall was warm, too. Citizens were happy, "Oh, how nice! October is like August!" But Chubais (*Influential member of Yeltsin's administration and the head of the Russian Unified Energy System—Translator's note*) nevertheless cautioned, just in case: "I still warn you that if temperature in winter holds around -25 degrees (-13°F) for more than three days, we'll be cutting off some units. Yeah, we'll cut them off. We will . . ." Luzhkov, Moscow mayor, was indignant: "How's that? The hard freeze—and he'll cut off, the villain!"

And so, in mid-January meteorologists told us that temperature would drop to minus thirty (-22°F), be ready. Well, unpleasant forecasts usually come true—the temperature did drop. And quite fast. In the morning, hardly awakened, I looked at the thermometer—it was zero (32°F). By evening, it dropped to -17°C (+1°F). And the next morning the thermometer was showing off with -22°C (-8°F).

"Will my car start?" wondered my wife.

"It should," I answered. "But just before you start it, turn on the headlights high beam to heat up the battery a bit. But don't forget that high beam works only when ignition is on. In other words, the lights on the control panel should be on . . . In about one minute you should turn off the headlights and turn the key back so as the lights on the panel would go off. Otherwise, the immobilizer will go off. It always goes off when the key is in the ignition position without starting the engine. And then you can start.

When it will start, let it warm up for at least three-four minutes, or better five. And then you should certainly heat up the automatic transmission. Press on the brakes and put the selector first in R for a minute, and then in D for a minute. You'll feel how the engine speed will go down. So don't start shifting the lever right away! If you'll be heating up the box right after starting the engine, the engine will simply stall."

"Oh gosh, it's all so complicated! Can't I just get into the car and start it as usual?"

"No, baby. The temperature dropped so much that we transited from the mode 'ordinary life' to the mode 'struggle for survival.' Why do you think the cold is the only news you can find on TV?"

Indeed, all the news programs told Muscovites that hard freeze fell upon them. In recent years, many new people bought cars, and after problem-free driving their leased foreign-made cars in several warm winters, these Russian novice auto-motorists for the first time have encountered problems with starting the car because of low temperatures. TV advised forgetful citizens that they were living in Russia and showed them mustached old auto-motorists who shared their old secrets of starting a car under abnormally low temperatures.

Turn on the headlights.

Bring the battery home for the night.

And, of course, disengage the clutch—those who have it . . .

Oil had to be changed for winter oil, instead of pouring in all kinds of all-season junk—hey, nitwit, it ought to be thinner. At worst, you can pour a cup of gasoline or kerosene into oil in the evening to thin it . . .

And northerners allegedly have pre-start heaters installed . . .

Besides, all mass media and Internet announced that emergency teams were established in cities and towns to be on duty round the clock—just in case. And as it is known, cases can vary. Metal can become brittle at low temperatures. That was the reason why the first day of the freeze in Vladimir province, a large-diameter water main burst, which supplied the city with water from the water reservoir. So what's next?

And all this anxiety with not even too low a temperature of -20 degrees (-4°F).

"And tomorrow it's going to be thirty (-22°F), honey."

At -20 (-4°F), the car started at first try, although the starter was, of course, turning listlessly and lights on the control panel were treacherously going off.

The next morning, I saw -30°C (-22°F) on the thermometer. I found in the attic and attached the underlining and the fur collar to my jacket, and used my old hat with earflaps to cover my head. The code-lock in the house entrance didn't function, nor did the door-closing mechanism. The cold immediately started biting my face and legs, easily penetrating through thick winter lined pants. I felt how tears, squeezed out by the cold, froze in the corners of my eyes, and how the eyelashes were freezing together. I don't know how the tip of my tongue touched the jacket zipper closed to the top, and screamed—the tongue froze to it! The hoarfrosted car, of course, didn't start.

"You should have removed the battery," reproached my wife. "You're old and experienced, and I'm young and inexperienced. Therefore, it's your fault. Why didn't you remove the battery last night? And why do they generally bring it home for the night?"

"For the reason," I tried to use the erudite theory to cover my feeble practice, "that battery capacity drops badly in cold temperature, so it doesn't have enough energy. But spending the night at home wouldn't help the battery a lot. The car fails to start not just because of the weakened battery. It can also have its oil frozen, which becomes like resin. Try to turn something in it, especially with a dead battery! Also, the gasoline evaporates very badly in freezing temperatures. And if there's no gasoline vapor in cylinders, there's nothing to burn! Plus the dead battery produces very weak sparks. So, one hundred different reasons—and all because of the cold."

"It's a shame my car didn't start! The neighbor said the entire city of Moscow is empty today—best time to drive. And the subway is probably overcrowded . . . The eyes are watering from the cold. Das ist terrible!"

Indeed, the cold severely disorganizes life. It puts on alert all city services, emergency, and transport.

The freeze from -20 to -30°C (from -4 to -22°F) stayed for almost two weeks. Newspapers, TV news and headlines reminded of news from the war front. In Orekhovo-Zuyevo (about 100 kilometers [60 miles] east of Moscow), the severe cold caused the highway bridge-supports to crack. Specialists from the chief office of the Ministry of Emergency Situations (MES) examined the seriousness of the crack and the capability of the bridge to stand. The Transportation Safety Inspection of Moscow Province was urgently studying the possible detours. But they had their hands full even without this. In previous days, transport police had to tow away 186 cars that stalled right on the road because of the hard freeze. In less populated north and east provinces, authorities and MES organized mobile rescue teams in vans. They drove along highways in search of stalled cars in order to save drivers from certain death. Traffic there is low and populated areas are hundreds of kilometers (miles) away, and while it's OK in summer, this time people could freeze to death waiting for a passing car . . .

Moscow Provincial Anti-Crisis Center deployed a reserve autonomous communication system between municipal authorities and emergency services of various agencies. Round the clock duties were organized at all the essential key points, power plants, and communities. In Moscow city and province, an emergency regime of energy consumption was introduced, under which a whole number of industrial enterprises were cut off from power supply. All construction works were stopped. Some of the trading areas and shops were closed. People stopped going to work, keeping warm at homes. In other words, the entire economy of the region was stifled, and all the freed power reserves were used to save people from ferocious death. In all the cities within 50 kilometers (30 miles) around Moscow, all electric transport was replaced with motorbuses. And in spite of all, power plants were working in the over-the-limit mode. Moscow devoured sixteen thousand megawatts of energy a day—and that with nearly all industrial enterprises stopped. Everything was spent on heating.

Vice Governor tried to calm down the population by saying that in case of crisis situations, four mobile units are deployed with mobile power plants capable of supplying power to a city of one hundred thousand population. Although, for not too long—just as long as the fuel would last.

In spite of all the efforts, bursts were happening all over the country—from Moscow to the Pacific Ocean. The evening news reminded the reports of Sovinformburo (the Soviet Information Bureau, established during World War II). The TV broadcast daily about the progress in various parts of the country in struggling with burst pipes, about saving residents from death by evacuating them, about heat guns installed in house entrances to prevent pipes from bursting in houses and whole blocks that were left without heating.

I was reading daily casualties reports on Internet: "According to ambulance service yesterday, 7 people died, 25 hospitalized for hypothermia, 93 called medics for frostbites and 44 of them also hospitalized . . ."

This war lasted two weeks. However, the temperatures around -30°C (-22°F) lasted only two days and nights. The rest of the time thermometers were floating between -20 (-4°F) and -25°C (-13°F). It's not too cold, you would agree. But the country has cracked. And what if the temperature were to drop to -35°C (-31°F) and hold for a week? The probability of accidents starts growing exponentially with falling temperatures. If the number of accidents and catastrophes were to increase by one order of magnitude, the repair teams would not be able to do all the fixes in time.

And if the temperature were to drop another ten degrees (18°F) and last as long, the country would be no more. And by the way, many climatologists predict this particular development—sharp cooling as a direct consequence of global warming. And they have weighty arguments for that.

CHAPTER 1

Do Not Give Them a Fulcrum!

Do you remember the movie *The Day After Tomorrow*? It was born out of these theories.

Astronomers are quite right: a lengthy period of reduced solar activity awaits us, the Maunder type. The attentive reader probably remembers that horrible time, 1645-1715, Maunder Minimum of solar activity. Our luminary reduced its heat emission by a miniscule fraction of a percent. It caused the drop of global average temperature by a half degree (by 1°F). And this killed up to a half of the population of Northern Europe.

Famine

True, the postindustrial civilization is more independent of nature than the agrarian was. But we still get our food products from agriculture—we don't produce them in factories. True, the modern-day agriculture has some reserves to grow. But these reserves of 10-20 or even 30 percent would hardly compensate for the total crop failures caused by such a catastrophic drop of the global average temperature as 0.5°C (1°F).

I'll remind you a picture: the end of the "velvet" season, the very time when Russian and German seniors like to get to the sea, because it's not hot, the prices have dropped significantly, and one can freely swim and walk under the "velvet" warmth along the beach, get suntan, and breathe the sea air . . . So, at this very time (late September—early October) in the Maunder Minimum era, the elders would not even think about going to the sea—except, perhaps, for ice-skating. I have already mentioned that the siege of Azov by Peter I, "the Great," in 1696 failed only because the entire Black Sea area was buried on October 1 under deep snow.

It wasn't Peter the Great who killed a quarter of Russia's population by his cruel rule, which he was later often accused of by descendants. The cold of the Maunder Minimum of that time and the crop failures caused by it were the culprits that mowed down Russia worse than plague.

The next Maunder Minimum starts in 2020. And it should last until the end of the century . . . Have you stocked up with non-perishable food, matches, and rounds of ammunition? . . .

Another thing should also not be forgotten—that troubles never come singly. Fluctuations in solar activity correlate with location of heavy planets that in turn provoke movements in the Earth's crust causing earthquakes and volcano eruptions. The more eruptions, the greater probability of a large one, climatically significant. We've discussed it earlier. We also know that after a volcanically calm period, a very active time is soon to follow.

The cluster eruptions would be the ones to serve as the little trigger stone that would start the avalanche of a great glaciation. And speaking about the latter, I refer the reader to the graph of global temperature fluctuations in the last nearly half million years (see fig. 1).

As you see, we're living in the very end of a brief warming period, which should be followed by a drastic temperature drop. Everything seems to come together. Our planet is more used to glaciations than to warmth: of the last 2 million years, 1.8 million were glacial eras. And if we take a similar graph for the last 3 million years, we would detect a general cooling trend. The world is sliding toward some Super Ice Age. It's very disturbing.

Look once more at the trembling in the area near the 0 date. And consider: if these tiny fluctuations of the curve, equal to a half or a whole degree (1-2°F), could shake the empires and lead to the death of half of the population in some areas of the planet, then what would be the scale of a catastrophe resulting from the temperature drop down to the Ice Age level—by 8-9 degrees (by 14-16°F)?

During the peak of the last glaciation, an ice sheet of half kilometer (1,600 feet) thick was in the place of today's Moscow, and mammoths and woolly rhinoceros walked in Crimean, Italian, and Spanish tundra.

No One Will Be Here, Thank You All, All Free to Go . . .

And to bring it all to one pile, I can't avoid showing one more graph—the temperature history for the last twelve thousand years (fig. 12). It clearly shows the general cooling trend that started taking shape in the last sixty-five hundred years. As you see, it also is inauspicious. Is the temperature dropping?

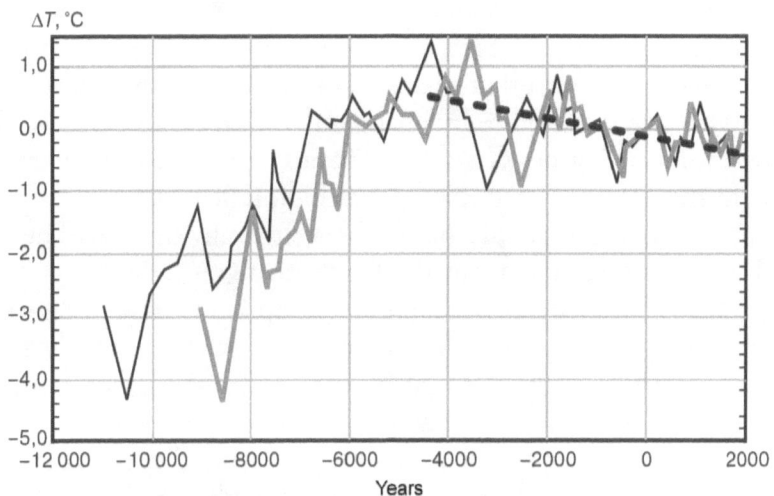

Fig. 12. Temperature history in the Northern Hemisphere during the Holocene (deviation from the norm in 1951-1980): calendar dating; radiocarbon dating; trend for the last 6,500 years

If so, then is it possible that all the talks about global warming are just a myth? Then what was the purpose for adopting the Kyoto Protocol? Why did academician Yuri Izrael advise Putin on radical ways to fight global warming? They were in fact so radical that it would be just a snap to reduce the crop capacity in Russia and in the entire Northern Hemisphere. This uncle suggested to drop the planet's temperature by two degrees (4°F) at one go! With a cavalry attack—simply to pump about 600 thousand tons of sulfuric aerosol into the upper atmosphere. How? Well, for instance, by using sulfuric fuel for several years in specially designed airliners.

. . . The effect of a volcano eruption, caused artificially . . .

And at the same time, the problem of falling crop yield, the possibility to move the climate system into a new glaciation, or the question of whether aviation engines would welcome the fuel with increased content of sulfur, were simply not taken into account. Because if they were, would the entire fleet of high-altitude airplanes have to be reequipped with new engines? Academician Izrael did not make it up all himself. He just explained in a new way a suggestion by another academician Mikhail Budyko, who in the time when everyone "was born to make tale a true story" (*quote from a popular Soviet song—Translator's note*), came up with this solution—to disperse sulfuric aerosols in the troposphere. A similar solution was suggested quite independently of Izrael and Budyko by a Moscow professor, Nurbey Gulia. Only he proposed to disperse aluminum powder in the upper atmosphere—it has high reflectivity.

In other countries, people are no fools either, by the way. They also started creating ideas on man-made climate change—although godlessly dittoing them from the Russians. For example, the *Acta Astronautica* magazine, the publication of the International Astronautics Academy, published an article on how to put tropics under shade and make the climate milder. To achieve that, a ring of dust should be created above the equator by carrying crushed dirt in rockets and dispersing it in the near outer space. The particles reflecting the sunlight could be taken from mines on either Earth or Moon.

A whole group of scientists working on the project under the manager Jerome Pearson, the president of Star Technology and Research, Inc., asserts that reduction of sunlight by 1.6 percent compensates for increase of temperature by 1.75°C (3.15°F). The Earth with such a ring would look like Saturn, and the dust ring would not only shade the tropics at day, thus softening the climate on the equator, but would also illuminate Earth at night. The cost of the project was up to $200 billion. A trifle—but nice.

. . . All rushed with fervor to save from warming the planet that's on the verge of great glaciation! . . .

The mentioned Professor Gulia has a whole sea of such ideas. For instance, how do you like this one: to change the angle of the Earth's axial tilt? It's by the way technically feasible, although somewhat too expensive. And too long. It would require building a circular railway tunnel with a radius of one-two thousand kilometers (600-1,200 miles)—just the size of Siberia—and pump out all the air from it to reduce the drag. Then a circular train filled with ballast should start running. In about five hundred to one thousand years, the Earth's axis would change its tilt due to the created angular momentum, which would cause the Earth's axis to precess. Once this Cyclopean structure is correctly designed and built, the Earth's axis could be tilted in any direction: for instance, the North Pole could be placed in the Pacific Ocean. In the latter case, both polar caps would be located in the oceans, providing for people a whole new continent—Antarctica—that would be located at the equator.

However, there's some unpleasantness here—the polar ice caps would melt, causing the World Ocean level to rise and flood some countries, including the entire Great Britain, half of Europe, St. Petersburg, all coral islands, part of the United States . . . But on the other hand, firstly, when the ice caps would refreeze at new locations, the water would subside and we would regain our Europe. And secondly, the flooding could be generally avoided if the Earth's axis were to be shifted slowly—so ice caps would move to their new locations gradually.

Or there could be a different way to manage the planet—simply to "straighten up" the Earth's axis, thus "eliminating" the change of seasons. Then on the equator always would be summer, a little away would be either spring or autumn, and at the poles would be eternal winter.

You might not believe it, but this project got even a copyright certificate #783520. And this certificate was issued back in the Soviet times, i.e., before the current total democratization of patent business allowing any idiot to receive a patent for any junk in exchange for money. In the Soviet times, the patents were based on the state rigid foundation; they were called copyright certificates and were issued not to anybody and only upon the most thorough scientific and technical evaluation.

It should be however said in all fairness that the Gulia thing was invented and copyrighted not to shift the Earth's axis but to accumulate energy—a kind of a super-flywheel to level the loads in the world energy system. Such flywheel could store up so much of the funky joules of energy that it would be possible to forget about all those peak-sleek loads. The unsupported "cars" with ballast are accelerated by the traveling magnetic field that uses the same principle used in all asynchronous electric motors, which by the way are the most widespread in technology. And since there's cosmic vacuum in the tunnel, the train can reach enormous speeds without using brakes. However, accelerating this "train" would require the amount of electric power that the whole world does not yet have! But there's no rush—for a thing like that, nuclear power plants could be specifically built to start the super-accumulator and then accelerate it in the years to come. The main point is to accelerate it only once, and then the circular train would run on its own for many years ... But if some catastrophe were to happen (for instance, terrorists blow up the tunnel wall) and all the energy of this super-flywheel were to be released at once ... Mamma dear! The planet would shake and half of Siberia would be destroyed ...

Gulia had another version of solving the problem of shifting the Earth's axis. Why build a super-pooper flywheel in Siberia, when the Gulf Stream could be circled? This would also create the angular momentum necessary for slow drift of the Earth's axis in the desired direction. All is very simple: we turn the Gulf Stream southward, so it now rounds Africa, flows into the Indian Ocean, and then, by way of the huge canals pierced through Himalayas with the use of 100-megaton hydrogen bombs, rushes to the Arctic Ocean through the widened riverbeds of the Ob, the Yenisei, and the Kolyma. The second branch of the Gulf Stream, getting past Cape Horn and Tierra del Fuego to the Pacific Ocean, also flows into the Arctic Ocean through the Bering Strait. Of course, in order to allow those masses of water to pass, the Bering Strait would have to be widened at the expense of a part of Chukchi Peninsula that's good for nothing anyway.

The cold water of the overflowing Arctic Ocean would then inevitably rush into the Atlantic Ocean past the coastlines of Western Europe, while on its way worsening the climate for those damn capitalists. Thus, we have succeeded in turning the Gulf Stream into a circular Pan-Oceanic stream. The important thing here is to start the "engine." And then, according to the inventor, this global stream would get into its "regime," becoming dozen times more productive than the today's Gulf Stream and at the same time supporting itself.

Let's do some calculations. The Gulf Stream transports about twenty-five million m^3 (6.6 billion gallons) of water per second. Considering that Pan-Oceanic stream would transport ten times more and the speed would be three times higher than the Gulf Stream, we come up with 10^{20} kg ($26 \cdot 10^{18}$ gallons) of water that would be in circulation. It equals to 100 million km^3 (24 million cubic miles) of water, which is not much, compared to the total volume of all oceans. To determine the moment of inertia of this water mass, we have to multiply it times the Earth's radius, which is 6,370 km (3,960 miles), and we get 10^{23} kg $\cdot m^2$

$(24 \cdot 10^{23}$ lb. $\cdot$ ft.^{2}). And to obtain the basic parameter enabling to shift the Earth's rotation axis—the angular momentum—we should multiply this value by the angular speed of rotation of this water about the Earth's axis. It's very small and equals $1.57 \cdot 10^{-6}$ radians per second. Resulting angular momentum of the stream is $6.28 \cdot 10^{27}$ kg$\cdot$m^2/s $(1.49 \cdot 10^{29}$ lb.$\cdot$ft.^{2}/s). But our planet's angular momentum, according to Gulia, is 10^{33} kg$\cdot$m^2/s $(2.4 \cdot 10^{34}$ lb. $\cdot$ ft.^{2}/s), or greater by six orders of magnitude. Nevertheless, a gyroscopic moment would occur, causing Earth to precess at an angular speed of approximately 10^{-9} radians per second. So, the professor estimates, in less than 10^9 seconds, or thirty years, we could shift the Earth's axis by ninety degrees.

I think if we were to shift the Earth's axis and to place it somewhere in the territory of the United States, the geopolitical Eurasian Mr. Dugin would be in total raptures and make something similar to the following profound statement in the TV program *Russian Home*: "The Lord gives us this unique chance!" It's hard to overestimate the geopolitical benefits that Russia would get from "freezing" the USA—the main Atlantic competitor of countries of Eurasian axis. The point is not just that Russia, already free from the idiotic democratic crap, could be called the birthplace of elephants. And by the way, settling these large mammals in the Russian Plain would not just significantly help the peasant labor, but fully once and for all provide the country with meat and milk (elephant milk is very nutritious and healthy). Besides, it would help the agrarian sector to save on metals and fuel, since in farming it's more beneficial to use herbivore animals as construction cranes and draft force instead of costly equipment that we can't properly make anyway . . . The main advantage of this geopolitical project is that the world would once and for all become unipolar, eliminating the main competitor who makes commodities of better quality than we do, thus undermining our economy. The Russian consumer, forced to buy homemade goods, would support our manufacturers. Of course, manufacturers of warm clothing would suffer huge losses, but on the country's large scale of things they would be overbalanced by profits from supplies of bananas to Africa and Southeast Asia.

OK, enough jokes. Here's another climate project for you—this time, quite a serious one, developed by two Russian scientists. The project is so interesting that, perhaps, I should devote a separate chapter to it.

CHAPTER 2

Saharan Sand

(The chapter title is a pun in Russian, when adjective Saharan *is the same as "sugary," and the term* sugary *[or* Saharan*] sand means "ground sugar"—Translator's note.)*

Do you know what Sahara is? A sandy desert with an intrepid Bedouin walking there . . .

Sahara is the world's largest desert, equal in area to the United States. In some days, the air gets as hot as in a Finnish sauna: 70-80°C (160-180°F). You could walk thousands of kilometers in Sahara and see no sources of water. Thus, life in Sahara is impossible, resulting in the whole country lost for nothing! And what's even worse, Sahara grows, expanding its territory and giving geographers and climatologists a reason to call barren Sahara the continent's cancerous tumor. And Sahara "metastases" are found even on other continents: storms scatter sands over huge distances—Sahara sand dust was even found in the atmosphere of the Southeast coast of the United States and Sahara sand deposits were registered in some regions of England and even Sweden!

But is it possible to return Sahara to humans in its primordial state? It was mentioned numerous times in the course of this book that Sahara was once a fertile savanna with people plowing, pasturing, fishing . . . What a wonderful place it was!

Returning Sahara to humans is a noble task. Two Russian physicists involved in climate have solved it: Yevgeny Demin, Ph.D., and Dr. Victor

Kushin. They developed a technical design of man-made climate change for one single desert of Sahara.

For a disease to be cured, its cause must be understood. So what is the cause of Sahara's excessive dryness? The scientists gave the following answer: the droughts are caused by stable anticyclones. Air masses from the cold layers of troposphere drop to the ground and heat up locally, forming a slow-moving anticyclone. Under conditions of anticyclone, no clouds are formed, which is why Russians love anticyclones so much: clear sky and bright sun are much more attractive to them than the boring leaden sky. And as it is known, rain comes from clouds. No clouds—no rain, and nothing hinders the sun from heating Sahara. Temperature rises higher and higher. At the same time, the coming monsoonal rain-bringing clouds are blocked by a peculiar heat curtain. A similar pattern is observed in other droughty areas of our planet with the same predominantly descending airflows and relatively stable anticyclones.

It's a vicious circle: the heat forms a heat barrier preventing clouds from coming, and cloudlessness allows the sun to heat the desert unhindered, forming the heat barrier. Following from this it's clear: in order to bring the Sahara climate to norm, we must break this positive feedback. To do that, rains have to be launched. How? Some people of Central America and Equatorial Africa have long discovered that fires in prairies and savannas often lead to the formation of cumulus clouds, and they started burning grass and brush on purpose to provoke precipitation. They succeeded on some occasions: large fires create ascending airflows opposing the anticyclone's descending airflows. But this method is unacceptable in Sahara: nothing to burn there. Therefore, a principally different approach is needed.

So, the stable dry anticyclone is our obstacle, right? Then instead of the anticyclone, we would have to start a cyclone over North Africa! As it's known, cyclone and moisture are twin brothers. A cyclone is formed in nature when ascending vortical flows of moist air are sucked into the area of lower atmospheric pressure and cooled while ascending. Cumulus clouds are thus formed, bringing rain.

In the latter half of the twentieth century, technology was born to create powerful vertical airflows—meteotrons. Roughly put, meteotron is a huge burner directed upward. First meteotrons were built back in 1967. Scientists planned to use them to provoke rains for agriculture and studied effects of large fires on the atmosphere. And by the way, the easiest way to make meteotrons is by using worked-out aviation turbines.

So, we're placing worked-out turbojet aviation engines in Sahara and directing them into the sky. Water is fed into turbines by injection. The gas stream temperature at the nozzle exit is about 500-700°C (900-1300°F), so the water evaporates instantaneously and dissolves in the ascending flow.

Two questions:

Where to get water in the desert?

How to start rotating the ascending flows in order to form a cyclone?

First, the needed amount of water would be huge—several cubic meters (about a thousand gallons) per second. There are three ways of solving the "water problem": use of desalinated seawater (via pipeline from the Mediterranean Sea); water from artesian wells (hydrologists discovered a big water reservoir beneath Sahara); or a pipeline from the nearest rivers—the Nile and the Niger.

The second issue is even easier. If meteotrons were placed on platforms moving along a circular railway with a diameter 100-120 kilometers (60-75 miles) and airflows directed at an angle with the horizon, powerful vortical flows are formed. The ascending moist air reaches upper troposphere limits with temperatures below freezing, where moisture condenses into water drops with some of them turning into ice. Precipitation falls in the form of water and ice, cooling the air and irrigating the soil. The most important thing is to start this process, and then it would support itself without the need for meteotrons. Occurrence of a small cyclone with precipitation over an area of 10-15 thousand km² (4-6 thousand square miles) is capable of breaching the huge Sahara anticyclone like a shaped charge. This breach would open the way for abundant monsoon rains. As a result, in just several years the humans would get enormous areas suitable for life. Wouldn't it be wonderful?

If you are disturbed by the huge scale of the project and thus its cost, I can reassure you. Calculations made by the authors indicate that implementation of this project would require as little as 0.1 percent of all the moneys spent for the operation "Desert Storm" or the second war in Iraq.

The authors calculated not just the economic part of the project but lots of technical details as well. To save on fuel, a method of water preheating was devised—the water first gets into accumulating basins where it is separated from air by a polyethylene film (to prevent its evaporation). But the film does not prevent water from being heated, and it heats up almost to the boiling point. And this already hot water is pumped into meteotron injectors. Hence, the savings—no need to waste fuel for heating the water from +30 (86°F) to +100°C (212°F). No-cost recuperation from an external source . . .

The proof of project's feasibility comes from a recently discovered phenomenon: geologist I. Yanitsky and geophysicist E. Borodzich discovered the occurrence of a local cyclone in the center of a vast anticyclone. They observed this phenomenon in West Mongolia. A similar anomaly was later observed in the Caucasus, too. In both instances, immediately after the cyclone nucleus formed inside the anticyclone, strong rains poured onto the ground. So this kind of process does happen in nature. And the task for humans is to trigger this process in Sahara, give it a nudge. It's interesting to note that lightning and thunder would start before the rain: when the process is launched, friction between the rising warm moist air masses and the dry ones would facilitate electrization. So observers would see lightnings and hear their crackling and rumbles of thunder above the installation! Albeit small ones . . .

The reason why I pleased the reader with all these projects was not to cause the reader to drop into the armchair and to start smiling dreamily, impressed with the range of human genius. Oh no, not for that! However, before I reveal my secret scheme, I want to tell you about a couple of other large-scale climate projects. Our inventors generally love everything large. And not only they, but politicians, too, who occasionally are forced to see face to face some crazy inventor who by strange chance breaks through the wall of scientific reviewers. And then the politician, the son of his people, drops into a leather armchair and smiles dreamily . . . I recall, in 1991 in the Russian White House (seat of the Government) I saw with my own eyes a crazy project of superspeed transcontinental railway using magnetic levitation. It was supposed to go across the entire Russian tundra, cross the Bering Strait (in a tunnel), get to Alaska, and then to Canada and the continental United States. The title page of the project had the signature of the Chairman of the Supreme Soviet (legislature), Ruslan Khasbulatov, with his instruction: "Please consider. I think it's interesting."

But let's not digress from climate. In the secluded times, almost epic today, when population was shaking from great achievements, and battalions of Stalinist writers, graduating annually in numbers from the Literary Institute, were supporting the Party line in every way, the talented

Soviet writer Alexander Kazantsev wrote a thick book, *Polar Dream* (Northern Jetty), about how the Soviet people are prevailing over climate. They build a super-dam on the bottom of the Arctic Ocean along the entire Eurasia coastline (!) so the ice of the cold ocean would not hinder the Soviet ships to sail along the Northeast Passage.

As a child, I used to read this thick fantastic crap-epic that resembled an industrial novel. And I advise everyone to read it—to be imbued with the spirit of that time. (And then read Solzhenitsyn—so the impression of that time would be full.)

Similar plans were numerous. Klimenko told me that when he was still in high school he read in the 1959, children's encyclopedia about a project to block the Bering Strait with a dam. "I still remember up to this day the colored picture—airplanes fly, trains run, Lenin's statue stands near the dam, pointing a finger in the direction of America . . . And they planned to pump warm ocean water from the south with huge pumps, creating an artificial warm current. There was also another idea—to use huge pumps for moving warm Atlantic water along Russia's north—extending the Gulf Stream to melt the ice of the Arctic Ocean, to establish all-year-round sailing along the Northeast Passage (a.k.a. Northern Sea Route), to plant wheat in Siberia . . . And there was another project more feasible—to flood Siberia, making a man-made sea in place of Siberian Plain by blocking the Yenisei. Creating an artificial sea was supposed to soften the climate in Siberia . . . And there was yet another project—to turn the northern rivers of Siberia southward. I'm sure you remember that."

"Indeed, I do. They even started digging . . ."

This is where I'm leading to, my reader: not a single country in the world initiated so many fantastic giant projects on changing the climate, as did Russia. What could be the reason?

CHAPTER 3

The Reason Is All the Same . . .

The best minds of Russia realized long ago: the main problem of Russia is lack of heat and moisture. Mendeleyev, Obruchev, et al. dreamed of modifying Russian climate. Those dreams began turning into projects in the twentieth century when people were born "to make tale a true story" (or to get to concentration camps).

In the 1960s, the Russian Academy of Sciences held several international conferences dedicated to the issue of artificial change of Russia climate. The Soviet authorities were ready to invest billions for at least a tiny bit increase in annual average temperature. Or at least in just the moisture alone! But today, when everything is happening by itself and cost-free, Russian ecologists are pulling their hair: oh, let's fight global warming together with the entire civilized world!

And getting back to the question asked earlier—is it happening at all? And if yes, then when would it be replaced with an ice age fatal to humans?

Here are my answers in the order of the questions. Yes, warming is happening. Many scientists up to this day, by force of inertia, call the climate of mid-twentieth century "contemporary climate." But today's climate is totally different! In the latter half of the twentieth century, the global temperature rose 0.4°C (0.7°F). We have gotten used to the idea that it's very much. Remember Ovid and his piercing shrieks from iced-over Romania. It was just horror there, wasn't it? The point is that the Ovid-era climate was precisely the same as the climate of mid-twentieth century. In the era of Ovid and in the era of Stalin the Earth was much colder than it is today.

In the twentieth century (especially at its start), frosts of 35-40°C (negative 31-40°F) were not rare in Russia's northern provinces in March-April. Today it's impossible even to imagine that! And the warming was taking place literally in front of our own eyes! I personally remember how we used to have PT school lessons in March on skis. Now in March, children have their PT lessons in a gym. In the last twenty years, March became warmer in Moscow by 4°C (7°F). And in hundred years by even more. While the average winter temperature in Moscow in the end of the nineteenth century was -12°C (+10°F), it was -6°C (+21°F) in 1990s. And it should rise as much in the next hundred years when the average winter temperature in Moscow would be around zero Celsius (32°F).

Wherefrom do we know this? From the climate mathematical model. Not too many sophisticated climate models are in the world today. Only developed nations have them. One model is in Germany, one in England, one in Australia, one in Japan, one in China, one in Canada, and about five in the rich USA. These are so-called models of general circulation of atmosphere and ocean, which use the set of simultaneous differential equations to describe simultaneously the behavior of atmosphere, ocean, and forests . . . The number of models available in the whole world could be counted on fingers. They are so few because they are very expensive—a model could cost $1-2 billion, and one country usually can't afford more than one model, while the majority of countries don't have them at all. The only positive exception is the rich USA.

Here's a reasonable question: why a mathematical model costs so much, when it's merely a computer program? Because it can't be used on an ordinary computer. Extremely expensive supercomputers are needed here. Only a few of them are in the world. They are made for specific purpose to compute climate and weather; their main customers are meteorology and climatology organizations and defense agencies too, of course.

Governments pay such huge amounts of money for weather computations only because the forecasts pay for themselves. For example, Americans built a thousand-mile-long oil pipeline across Alaska. Bur prior to that, they needed a forecast of the permafrost behavior there with respect to global warming—would the soils hold or move causing the pipeline rupture. It would also be good to know how the climate change would affect the fish catch, the ecological tourism, the frequency of hurricanes and storms, etc. The climate forecast affects prices of coal, oil, metal, electric power, as well as sowing times and crop capacity ... Money depends on climate! That's why the climate forecast is so important. And not just for Americans.

Obviously, there are no supercomputers in Russia. Therefore, Klimenko had to solve the task of climate forecasts in a roundabout way. And the task was solved. And it was solved better and more accurate than abroad. Poverty occasionally helps, forces to think—the mechanism here is the same as during global cooling: deficiency of resources forces people to find their ways, to use their brains.

All known world climate models are: (a) purely theoretical; and (b) basically taking into account only the change of one climate factor considered major. Currently, this factor is considered increase of concentration of carbon dioxide and other greenhouse gases in atmosphere. But this factor is not the only one!

As I have already mentioned, Klimenko decided to use a different way that seems at first glance more complicated. Firstly, his model is empirical, that is, it's built on the basis of experimental data. This means the coefficients in formulas are not devised but taken from life. They have no theoretical explanation at all but rather have proof by practice. Secondly, this model is complex, taking simultaneously into consideration a number of factors: movement of heavy planets, behavior of Sun, position of Earth on the orbit, volcanic activity, amount of greenhouse gases, etc ... The laboratory has a unique collection of all major parameters affecting the climate.

I remember when once telling me about his model, Klimenko took several thick paperback books and threw them on the desk. "Data collection takes great efforts. Every month I read several books like these—here is information on oil and gas extraction, areas of arable land and bogs, total number of sheep, sugar-cane production ... And you remember that sheep are one of the main suppliers of methane to the atmosphere. One molecule of methane causes greenhouse effect 20 times stronger than one molecule of the notorious CO_2. There are billions head of cattle in the world, whose effect on the atmosphere is comparable to the effect of bogs that produce methane, too. In general, to make the computations correct, hundreds parameters should be taken into account. It's necessary to add up sheep, goats, and bogs including Siberian, American, and African.

Plus, Russian holey pipelines with methane leakages, and plus coal mines where methane often explodes . . .

"Methane in general is a very insidious thing. It's much easier to handle carbon dioxide that's supplied to the atmosphere mainly through the burning of fossil fuel. Thus, we can indicate the single main source and work with it. But methane has about ten sources and each "weighs" about 10 percent. So you can't ignore any one of them. And these are only two gases! In general, there are about thirty greenhouse components in the atmosphere! Freon in refrigerators, manufacture of polyurethane foam and new packing materials—the computation model needs all this to be monitored and the data entered into it . . ."

In spite of seemingly simple explanations, Klimenko model is by no means simple mathematically. It took twelve years to develop it and to accumulate the unique computational database. The model's development started back in those prehistoric times with 286-computers around (the youngsters, of course, don't remember this). So, it took several days for a 286-computer to run the first versions of the program. On a Pentium-base, it takes about an hour for computing one version "from pushbutton." "Computing one version" is, for example, climate forecast of winter seasons for Moscow region for several decades ahead.

The model's accuracy is astonishing. In 1994, Klimenko took a risk and published his first forecast of climate changes up to the year 2005. For average five—and ten-year values, the forecast came true with the accuracy up to 0.02°C (0.04°F), or the same accuracy the climatologists use to measure global and hemispheric average temperatures. You can't do better than that even if you wanted. Since 1996, laboratory made twelve successful forecasts for twelve seasons (winter, spring, summer, autumn) for Moscow region.

With the same incredible accuracy—up to one hundredth of a degree (0.02°F)—Klimenko laboratory forecasted planetary average climate for 1990s! The forecast predicted the deviation from the climate norm for the twentieth century of +0.4°C (+0.7°F), while the commonly accepted viewpoint of scientific climatologist community gave a prediction of plus 1-2°C (plus 2-4°F) from the norm. In other words, the world climate models, computed on supercomputers, gave an error by two to four times. Klimenko, using Pentium, hit the nail on the head.

By the way, making a forecast and waiting to see how turns out is not the only method of testing the model for its workability. It's possible to make a "forecast" for the past! Just run the model for "retro-prediction." And this model passes this test, too, providing perfect coincidence with paleoclimatic reconstruction.

As of today, Klimenko model has no equal for accuracy. As to being influential, Mr. Klimenko is far behind the Western "modelers" who tell lies godlessly and scare people shamelessly with the new World Flood. And it's easy to understand this paradox, knowing the principles of financing and functioning of Western science.

The scheme works as follows. A scientist makes a sensational statement. Sensation is news. And news, in accordance with all media laws, can't be good. Human and mass psychology is structured in a way that it perceives bad news with greater attention and interest than good news. Is it possible to call "news" a message that a house was built or nobody died in streets? Is it possible to call "news" a report on cutting ribbons at a bridge dedication? Or on the new products made at a factory? That's not news—it's like one half of worn-out jeans. So if a scientist makes a loud alarmist statement, he is paid heed and money for research. And if a scientist says that nothing bad would happen, then . . . what did he actually say? Who would give him money for research? And why he should get the money if the world is not threatened with anything?

Every year many various hypotheses appear in various fields of science. The media pick up those that (a) look sensational and (b) are akin to societal frame of mind. Today, green and ecology are in fashion. And the science in the West sits on the financial needle of grants. And distribution of grants is in the hands of people far from science, who might have heard something somewhere. The alarmist scientists feed the media with news. And the media feed the scientists: sensing where the media wind is blowing, dozens of researchers apply for grants, adjusting their applications for their maximum "topicality" according to the topics described in media. And naturally, research by the "granted" scientists fully meets the expectations and views of grantors. And the myth thus starts feeding and procreating itself.

The "modelers," happily computing catastrophic forecasts on their supercomputers, periodically predict that the global temperature would rise by 5°C (9°F) in the next hundred years, causing the new World Flood, the Gulf Stream to turn south, and Europe to freeze.

"I worked in the West with these people for many months and even years in the same institute in adjacent rooms," Klimenko complained. "My main complaint against them is that they don't use empirical information much, they are very keen in what they compute using their models. When I was telling them that their results did not correspond to the data from instrumental observations, their response was simply shocking to me. They told me literally: 'So much the worse for the observations.' These computer 'modelers' of climate are a separate social group living its own specific life that is far from reality."

What's dangerous about "modelers"? Their influence. Their influence facilitated adoption of the Kyoto Protocol threatening to stifle Russia's economy. And the Kyoto Protocol is not the first and probably not the last case when grant-fed alarmists move countries and continents to adopt senseless or simply harmful decisions.

And perhaps, I should dwell on this in some detail. Do forgive me for this digression . . .

Chapter 4

(Digression)
Legends and Myths of Our Time

Diseases can be real and imaginary, like for example, false pregnancy. And then there are sociopsychic diseases. I'll try to show several examples of these societal disorders, although I don't dare to assert their "falsity" or "realness": after all, even if the problem is far-fetched, the society suffers from it really. Some suffer, while others profit . . .

"Y2K Problem"

Do you recall? Oh yes! Because of the fact that in old times, primordial programmers referred to the year with just two digits, while new programs were made from ready blocks of old programs, the two-digit year format roamed into the future and up to the end of the century. And by that time, the world became totally computerized and all transport, financial, and life-support systems were dependent on computers that were about to go crazy on January 1, 2000, because of the "zeroing of time."

. . . Airplanes will crash. Lighting will turn off. Trains will derail. Reactors at nuclear power plants will get out of control and runaway. Nuclear warheads will break loose. Hospitals will give patients wrong medicines . . .

Those were the predictions. It was filmed in a Hollywood movie. Media wrote about it. Director of the International Y2K Cooperation Center, Mr. Bruce McConnell, was predicting in his report to the US

Senate that the consequence of the Y2K Problem would be a general recession of the economy.

The Head of the National Railway Administration of France, Mr. Louis Gallois, stated that they were working on solving the terrible problem since 1996, spent successfully 300 million francs, but had no confidence in successful outcome, and therefore all railway traffic in France would stop that night.

American computer companies spent several billion of budget dollars to solve the Y2K problem and were horrified: could those scanty sums of money prevent a national catastrophe? Naturally, the Russian legislature, the Duma, reacted to the global threat, too. Russian legislators react to everything every time in the same way: they write a legislative bill. In a matter of just half a year, the deputies wrote and passed a law on Y2K Problem. The rush was dictated by urgency: the homeland was in danger! Fortunately, the president vetoed the bill. Otherwise, our legislators would have saved the homeland—they can do it for sure . . .

The terror vanished with the start of the New Year. Like nothing ever happened. Nothing changed in this world when the ball dropped—trains didn't derail, airplanes didn't drop, water ran from faucets, power plants were producing uninterrupted electricity . . .

But just some money found its way into some pockets.

Ozone Hole!

In 1969, scientists for the first time registered the shrinking of the ozone layer over Northern Hemisphere. By that time, everyone knew that ozone protects us from hard ultraviolet radiation, which was why people were concerned. They started computing. It turned out that shrinking of the ozone layer by 1 percent would increase the number of skin cancers by 6 percent. Oh, what a disaster! What did we do to cause it? Is there a salvation?

Fortunately, in 1973, a German physicist Crutzen and two Americans, Rowland and Molina, discovered that chlorine-fluorine-carbon compounds (Freon), which are widely used in refrigeration and other industries, can start reacting with ozone and thus destroy it. This is true. But it's also true that ozone is a very active substance and can easily react with almost anything. But the media, sensitive to sensations, immediately trumpeted that women using sprays from aerosol bottles are destroying the ozone layer that's so valuable to humans. Media found the culprit and exposed it many times over! The German and Americans received the Nobel Prize for their wonderful discovery. Since that moment, there was no way back.

Socialist countries—Norway, Sweden, and Finland, immanently gravitating toward everything green, feministic, and egalitarian—suggested saving the planet from the contagion by total stop of Freon manufacture. The world began thinking—after all, the talk was about total restructuring of chemical industry. But in 1985, English scientists discovered very timely that ozone concentration over Antarctica was diminishing, too. Right then the Vienna Convention was passed on protecting the ozone layer and the famous Montreal Protocol was signed on limiting Freon production.

The USSR was then just at the start of its restructuring. The Red Bear tried its best to look as decent and civilized as it could, and therefore it got dressed in a folksy dress and danced a folksy dance on its hind legs. The Soviet Union tried to do everything at once for the sole purpose of being praised—it was pulling out its forces in record-short times and reducing its armaments. What else do you want—fight Freon? A piece of cake! In 1988, the USSR signed the Montreal Protocol and promised the civilized world to cut production of Freon by 50 percent in ten years. And in 1992, the renovated sovereign democratic Russia, while still in a state of freedom euphoria, signed in Copenhagen an even stronger version of the Montreal Protocol, according to which a total ban on Freon production would be introduced on January 1, 1996 (this, of course, was not fulfilled, and Russia publicly repented before the whole world, pretending being confused). By the way, the awakening China never signed anything: "When our economy would become as strong as in developed countries, then we shall all fight together for universal salvation. After all, the United States, and not China, produced the most amount of Freon, so it's for the United States to patch the ozone holes and not put the fight on the shoulders of less 'guilty' others."

A reasonable question: why was it the United States to fight Freon so fiercely? Very simple: the Montreal Protocol was signed only because the transnational chemical monster, DuPont concern, voted both hands for it.

For your reference, the world's largest chemical corporation, Du Pont de Nemours, was established in the United States in 1802 by a descendant from France whose name was Du Pont. By the time of our story, the corporation owned 225 plants on five continents. The value of DuPont goods sold annually worldwide was almost $40 billion. And DuPont invested annually about $2 billion in scientific research.

By the mid-1980s, the corporation was the world's largest producer of Freon, and the company obviously did not like the attacks against Freon. For that reason, DuPont spent huge amounts of money on debunking the Freon-ozone theory. However, the concern managers were no fools! They simultaneously invested huge sums on search for Freon replacement—just in case. And the billionaires saw their luck—they found a replacement for Freon and patented it. After that, DuPont became the world's most ardent defender of the ozone layer! Having the monopoly on the Freon replacement, DuPont was using politicians and greens to fight for worldwide ban on Freon in order to put underdeveloped and other countries on its chemical needle. Thus, the unproven scientific theory enabled the quick billionaires to earn a lot at the expense of the fools—in the United States alone the customers paid $220 billion to stop using Freon.

And what about the ozone layer? It was revealed later that ozone holes appear, as a rule, over the zones of tectonic faults. All ozone anomalies known today, when transferred from the maps of Central Aerological Observatory onto the map of geological faults, coincide with it perfectly! Depletion of ozone happens due to active degassing of Earth's crust that emits hydrogen and hydrogen-containing gases. They are the ozone destructors. And comparing Freon emission from aerosol bottles and refrigerators to planet's natural degassing is simply laughable—Earth emits ten thousand times, more gases than all the planet's chemical plants.

And what's interesting, the periodical thinning of the ozone layer followed by its thickening is our planet's natural process. It's as natural as fluctuations of the Caspian Sea level, which once also was a point of a fight. Research done by the associate member of the Russian Academy of Sciences Andrey Kapitsa and other scientists demonstrated convincingly that ozone holes occurred on Earth long before the first chemical factories appeared. And the reason is that the extent of Earth's crust degassing depends directly on solar activity. These two curves match perfectly.

SARS (Severe Acute Respiratory Syndrome)

Do you remember this scare several years ago? SARS is a scary deadly disease! It was the only thing talked about then. Borders were closed, quarantines established. People were removed from trains . . .

Prime Minister of Russia at a government meeting ordered members of the government to cut the number of passage points on the border. At airports and railway terminals, paranoid thoroughness was used to detect all coughing persons.

The SARS project developed in accordance with standard patterns. On March 14, 2003, the world media announced the new deadly disease in Asia, which is transmitted by aerosolized droplets. It must be a new virus sneaked up imperceptibly! Indeed—they found the villain as soon as March 19! They chose as culprit a virus from a family of coronavirus. The bastard is caught! However, several weeks later this virus was exonerated, and the "killer" title was given to a changed virus of common cold. A mutant!

New virus—it's always good. There are no antibiotics against it, no diagnostic instruments, no vaccines . . . Everything has to be developed anew. Give us the money! We have many experienced scientists ready to do the work and to save the humanity—for several billion dollars. Even in Russia, scientists began showing some activity—their hands were itching to save the humanity.

"You want money? I've got it!"—answered the international community. Hong Kong gave $26 million to fight the new enemy of humankind. China is not far behind, presenting $3 million monthly to its virus-fighting scientists. Japanese government bangs and throws $500 million to study the new virus. Who can do more?

But let's stop, cool down, and think—why such huge moneys are allocated for a semi-mythic disease that took away not many lives? And at the same time, the long-known diseases are taking away every year as many lives as the USSR had lost during the four years of its fight in World War II. Tuberculosis alone kills three million people every year. And it's not just TB! The common flu kills two million people annually. And how many people died of SARS? Just a few hundred? And maybe it wasn't even SARS—who knows after it's gone . . .

So the SARS story had the same ending as Y2K. Money was successfully spent and soon the Associated Press reported that WHO admitted that half of the Taiwanese patients were wrongly diagnosed with SARS. And the majority of SARS patients who died during the pandemic were seniors who had other viral diseases that were aggravated by SARS. And those complications were the direct cause of their death . . .

Avian Genocide

The punitive expeditions against chicken have rolled all over the world. Domestic birds were executed in huge numbers, demonstratively, and with good taste. The ditches filled with dead and burning bodies vaguely remind of something, but off with the doubts if the horrifying, deadly threat hangs over humans, which killed already . . . it's frightening even to say how many—maybe ten, or maybe fifty worldwide in several years. Hot on the heels of it, the malicious virus was revealed and shown to the public.

. . . Goddamn it, it vaguely reminds me of something . . .

Well, I'm not going to say anything about the bird flu myself—you got my opinion. Let's hear another one. So, who in Russia knows birds? Well, many do! For instance, Sergey Lisovsky, the senator, member of the Federation Council. I'll phone him. Firstly, he has a sober view of things. Secondly, why should I express my own opinion about chicken and viruses in a book about climate, when I can take cover behind an opinion of a respected person who knows chicken from A to Z?

"Sergey, bird flu . . ."

"Oh-oh," Sergey immediately was on. "A painful topic! Generally, the bird flu has been known for a long time; birds worldwide get sick with it periodically and die. For example, a heavy outbreak of bird flu happened in America in 1983, but nobody raised much noise about it then. Americans simply and quietly destroyed seventeen million birds then—and that was it. In order for you to understand the scale of things, seventeen million birds are not much—just four large factories. And the country has hundreds of them. But even after that case, once every two to three years, several US factories get the disease, and then they are being closed and the sick birds are destroyed. Without making any noise! And by the way, no humans died yet.

"In this connection, it's interesting what happened in Turkey where children died, allegedly from bird flu. Those children lived in slums, and children living in slums have thousands of causes from which they can die! So what, if the virus was found in blood? People have thousands of viruses of all kinds in their blood! Who could know that this was the killer virus and not the common flu virus that was in the blood, too? And generally, this whole story about the bird flu reminds me of a PR campaign. The time spent on this problem by the media does not correspond to its significance. CNN dedicates to bird flu half-hour of every two hours of its broadcast. It all looks like a well-coordinated provocation!

"This happened more than once already. What about the mad cow disease, with so much shouting and scare about it? But all that noise resulted in Europeans ceding their beef market to manufacturers from other countries with the price of beef jumping threefold! And by the way, another feature of similar PR campaigns is that they spread like fire. When it was related to Brits, the French were happily writing about mad cows, aiming at eliminating British beef from the market. But then the informational fire spread onto the continent and struck the French economy. The French wanted to gain by other's fire, but burned themselves. The same with the bird flu—first the fire burned in Southeast Asia, causing Europeans to feel happy, but then the fire spread into Europe. And now France, having learned the bitter lesson, allocated money for an explanatory campaign about bird flu being not dangerous for humans.

"I remember a story in this connection. In June 2005, before the bird flu outbreak, a delegation from the US Senate Agriculture Committee visited Russia to lobby for American meat interests—to increase Russia quotas on Bush's legs (*Russian term for chicken leg quarters imported from the United States—Translator's note*). The interesting thing here is that three members of the delegation were officially representing intelligence! I asked the Americans, why were they trying to push their chicken into our market, when our poultry industry was on the rise and in the next six to twelve months would be able to fully satisfy the needs of Russians in chicken meat. One of the Americans responded after a pause, looking straight into my eyes: 'I think this will not happen'. I didn't know at that time what he meant, but a couple of months later the bird flu story broke out. And it would be wrong to think that it didn't affect Russia. It certainly did! The wave of scares caused the consumption of chicken meat to drop 30 percent. The big factories would certainly survive, but the small and middle-size ones would not."

Chapter 5

Worldwide Floodlessness

Some climate modelers vituperate in press hysterically and demagogically: "The temperature will jump by 5-6 degrees (9-11°F)! The world will become so warm that all polar icecaps will disappear, glaciers will melt, and Europe will be flooded."

The others echo them: "Oh yes! Global warming is terribly dangerous. It will lead to a horrible cooling, because the melting Greenland ice will dilute the Gulf Stream that would start flowing in the opposite direction and cause Europe to freeze. All's very bad."

Many things happened in the long climate history of our planet. We had blooming warmth, we had ices advancing and covering up to half of Eurasia with huge polar icecaps, we had mountains grow and disappear ... But one thing never happened—the Gulf Stream to flow in the opposite direction.

Now as to the World Flood—let's look at the first article we find with a characteristic title "Arctic Ices Disappear" and see how they are trying to scare us in anticipation of grants and budget allocations: "Scientists from all over the world, working for the UN ecological program, sound the alarm in connection with the Arctic situation. They warn that the Arctic ice would completely melt in just sixty years. And then the life on Earth would change beyond recognition. According to the scientists, the worst to suffer from the melting of ice would be northern provinces of Canada, Alaska, and Russia."

The article says that 250 scientists and 6 international organizations worked on this report, warning the world of the severe danger. The work took four years. The money was successfully spent, and as a result—enjoy this horror movie.

. . . Arctic ices are melting. The level of the World Ocean is rising. Humans are leaving coastal areas in panic. Vacated are such cities as London, St. Petersburg, Marseilles, et al . . .

Professor Preobrazhensky once gave an advice to Dr. Bormenthal: "Before going to bed, don't read Soviet newspapers."

And I would add "And alarmist scientific studies, too."

Because if you do, you become the same kind of neurasthenics as Madam Nicole Soltman, manager of the program of climate changes of the World Wildlife Fund.

Ms. Soltman demands urgent reduction in carbon dioxide emissions and tries to scare everyone: "Life on Earth will change beyond recognition after the loss of the ice cover in the North Pole and the rise of water level of the seas, which will threaten the world's largest cities, such as London."

. . . How muddle-headed she must be . . .

Take a cup, fill it with water not quite to the top, and carefully place several ice-cubes so the water would reach the top of the cup. Add more water if needed, and if some would spill out—it's OK.

So, we're having the water level of the World Ocean with Arctic ice. At our cup's edge is London. Since it's warm in the room, ice starts melting. Here's a question for fifth-graders: when all ice is melted, would the water level in the cup rise? Would it spill over the edge and thus flood London?

The correct answer is "No." The water level in the cup would stay the same and would not burst its banks. Why should it? The ice in the cup (as well as in the Arctic Ocean) is afloat and in the water! So what's the difference whether this ocean water is in a liquid or solid state?

. . . It's interesting whether Madam Nicole Soltman ever drank whisky with ice? Let's say, she hasn't. But what about the other "scientists" who are saying the same thing? . . .

There is however a nuance. The ice placed in water has 90 percent of it submerged and only a small piece above water. So could this "extra" above-water ice cause some flooding? No, it can't, because the ice density is less than the water density. And when ice is melted, its volume is reduced by exactly the above-water piece.

The alarmists are right in one thing: the iciness of northern seas is indeed falling—at an approximate rate of 1-1.5 percent a year. Therefore, within 100-150 years the total destruction of multi-year pack ice will happen. And the Arctic Ocean will no longer have a solid ice cover it has today. We shall discuss later the benefits this would bring to Russia, but let's get back now to the flood.

We haven't yet mentioned Greenland, Antarctica, and mountain glaciers, where ice is not floating. It rests on dry land, but if melted it would honestly flow into the World Ocean. Well, we dare to ignore the mountain glaciers—they contain just 1 percent of the world's ice water. Whereas, 99 percent of continental ice is deposited in only two ice sheets—90 percent in Antarctica and 9 percent in Greenland. And Greenland has glacier pieces periodically breaking off and spreading as icebergs in Northern Atlantic. But this is nothing new, with icebergs in the past even hitting some Titanic et al. It's a normal phenomenon. Glaciers should normally slide down into sea.

Every day, twenty million tons of ice slide down from just one Greenland glacier, Jakobshavn, and float away into the ocean. And this process is thousands years old. People wore skins, invented iron, built the *Titanic*, and the ice continued to slide and slide down . . . And then, one wonderful day, some prince of Monaco or some great actress-protector of animals wearing an artificial fur coat visit Greenland on a tour. They open their mouth and make a loud statement for press—global warming should be stopped urgently: look what's happening!

Could the Greenland and Antarctica icecaps melt? No, they can't. As a reference for panicking ecologists, ice melts at a temperature above 0°C (32°F). Annual average temperature in Greenland and Antarctica is about -15°C (+5°F). In the last half-century, temperature in Antarctica increased by 2.5°C (4.5°F). According to the most catastrophic (and as catastrophically incorrect) forecasts, global warming could cause the temperature to increase by another 5 degrees (9°F).

-15°C + 5°C = -10°C (or 5°F + 9°F = 14°F).

Minus 10 degrees (+14°F)! It's still too far from the melting point, isn't it?

Above all, Greenland is a mountainous country with an intricate topography: mountains are higher at the edges of the island than in the middle. Even if the ice were to melt, it would result in a lake. But there would be no lake, because the world climate system is so structured that at times of global warming Greenland gets colder! Global warming by 0.6°C (1.1°F) does not imply that average temperature rises everywhere

—

by the same value. For instance, somewhere in Russia it could have risen by 6°C (11°F) (we'll talk about it later), while in some other place it could have even fallen—just like in Greenland. The same happens with precipitation areas. Global warming causes increased water evaporation from oceans and increased average total precipitation. But that's only in average! At various places, somewhere precipitation could increase a lot (like in Russia!), while other places could experience drought (like Africa).

Speaking about precipitation—observations show that ice thickness in Greenland and Antarctica grows! In the last thirty years, according to satellite monitoring, iciness of seas around Antarctica increases at a rate of 1.3 percent or 140 thousand km² (54 thousand square miles) in a decade!

The mechanism of this phenomenon is clear: global warming increases water evaporation from oceans. And then this water drops in the form of precipitation all around the world, including Greenland and Antarctica. The more precipitation—the thicker the icecap.

But then why semiliterate western climatologists and totally illiterate ecologists are screaming that the Antarctic icecap is crumbling? Because it is crumbling! But only just partially. For example, in West Antarctica, which geographically is an archipelago covered with ice but climatically differs greatly from East Antarctica, the shelf ice is indeed crumbling rapidly. First, the Larsen-B shelf collapsed on the western coast of the Antarctic Peninsula, and then, to the great surprise of scientists, the Larsen-A shelf, too, collapsed in two months.

But what is shelf ice? Here's the classical definition from a dictionary: "Shelf ice is a glacier that is afloat or partially resting upon sea bottom, which flows away from the coast. It appears like a plate with a precipice at its edge. Shelf ice is supplied from inflow of continental ice sheets, accumulation of precipitation, freezing over of seawater, and attachment of icebergs. Shelf ice is spread mainly around the coasts of Antarctica."

That is, it's almost the same as the ice at the North Pole—ice is mostly already in water and thus having little effect on the rise of the World Ocean water level.

But why are they crumbling, after all? Because this piece of Antarctica represents a warm oasis that is much different from the rest of the continent. It even has two kinds of flowers growing there! So why be surprised that this flower garden is affected by global warming?

Therefore, the panic about the World Flood is the destiny of the greens, social democrats, and idiots. The Flood will not happen. But the World Ocean water level will slightly rise. Why would it rise, if Greenland and Antarctica have their icecaps being frozen over and accumulating world water? Well, because with the rise of temperature all bodies expand (see "natural science" for middle school). And the body of the World Ocean expands, too, and thus the World Ocean water level would rise—no more than 1/2 millimeter (0.02") a year, or 5 centimeters (2") in one hundred years. No one would even notice.

And besides all, the very thick ice sheets of Greenland and Antarctica have colossal heat inertia. And it's only natural: the thickness of the Greenland ice sheet in its center is about 3 kilometers (almost 2 miles), and thickness of Antarctica ice sheet is more than 4 kilometers (2.5 miles). It takes a heat signal six thousand years to pass through those depths. The beds of these sheets still "don't know" about Julius Caesar or Egypt's pharaohs: the heat wave of that era has not yet reached the foundation. In other words, even if temperature rose to 30°C (86°F), it would have to hold for at least a thousand years for Greenland and Antarctica sheets to start crumbling significantly. So, we'll speak no more of the World Flood. Let's rather speak of the break that united almost the entire civilized world in the fight against global warming. And most active excesses on this front are committed by Europe.

I mentioned already that the winter when this book was written was a cold one. And it was cold not only in Russia. All of Europe moaned with abundant snowfalls and freeze. But in spite of the moans, Europe remained Europe after all . . .

In late January, my wife and I went to the West for cross-country skiing. We went not to "real" Europe, but to East Europe that's closest to Russia—Bulgaria. After hiking, we rented a car and toured the neighborhoods. We visited an ostrich farm, saw ostriches in pens, and moved to the adjacent enclosure. We saw some deer there. What marvelous darlings! They gathered at the fence, waiting for some kind uncle from Russia to give them a treat. The kind uncle shed some fond tears, bent down, picked some fresh grass, and passed it through the fence rods. Deer began to chew the food methodically. I bent down to pick some more and just then realized what was happening. I was picking fresh-grown grass in mid-January! That's what Europe is . . .

I wish we had such a climate.

Well, it seems these things are improving in Russia, too. And all thanks to global warming. Just recently, Moscow Botanical Garden workers discovered apricot trees in Moscow. The first tree was found in early 1990s in Pererva district. Seeing this miracle, the botanists couldn't believe their eyes—this never was before! They decided it was just by chance: "Perhaps, special microclimate is here—river bend, heating main . . ."

But it wasn't just a random chance. Today, jolly yellow apricots could be seen in many parts of Moscow. Even individual homeowners (dacha owners) in Moscow suburbs started planting them. Peaches, too, appeared in Moscow. The first peach tree was planted in the Textilschiki district, and the second near the "Serp I Molot" (hammer and sickle) works. The first fruits from those trees, tested by scientists, showed that peaches at the former site were dry and not sweet, while the latter ones were juicy and sweet.

Obviously, nobody planted these trees—just someone walked along a street, ate a peach, and tossed the stone. Or someone was riding in a commuter train and spitting out the window. In an area around a commuter railway, someone seems to have spat successfully with Isabella grapes—a vine grew up there since. Moscow still has not enough summer heat for the vine to produce berries, but it blooms quite southerly. Also, in the area of a cargo railway station, scientists discovered a whole thicket of quince.

The reason for all this is clear—global warming. Just about three decades ago, not a single southern tree could survive in Moscow—they all froze in winter. And today they do survive, with winters in Central Russia becoming mild and humid.

Is it bad?

Two primary troubles of Russia—they are not the fools and the roads (these are derivatives), but rather the cold weather and the low moisture—seem to start retreating. So there's no more need to block the Yenisei and create man-made seas all over Siberia, no more need to pump up the much-cooled Gulf Stream into the White Sea, no more need to build the Northern Jetty as designed by the writer Kazantsev . . . What's needed is simply to open the mouth and to consume the manna pouring down from heaven.

On July 28, 1601, "heavy snow fell in Moscow amid summer and it was freezing and people rode in sleds," while the country had nearly died out from famine ("huge numbers of starving Christians buried"). In

July 2005, Moscow dacha owners stocked up jars in anticipation of high harvest of apricots. Which do you like better?

Russia is a country of permanent climate catastrophes: in the last one thousand years, Russia had 433 years of dearth. Do you still want to fight the global warming? Do you still need the Kyoto Protocol?

CHAPTER 6

All Join in Fight Against Warming!

The PR project "Global Warming—Nightmare, Nightmare!" was developing in accordance with the same scheme as the other alarmist projects—"Y2K," "Freon Destroys the Ozone Layer" . . . Sensational information in media with reference to scientists—public concern—allocation of money—justification of spent money by new loud statements—panic—adoption of international resolutions—new money.

Upon learning firmly from media that greenhouse gases contribute to warming, politicians of a number of countries invented an easy recipe to fight the warming—it's necessary to emit less greenhouse gases into the atmosphere. But who emits gases? Industry, power engineering. An international agreement was composed urgently, according to which the developed countries voluntarily agreed to limit emission of greenhouse gases and primarily CO_2. In other words, to stifle own industry and power engineering.

In December 1997 in a Japanese city of Kyoto, this document was signed on a wave of fear by a number of countries. Among those intelligent signatories of social-democratic brand was one brown bear—Russia. As you remember, Russia was then in that kind of a gloomy condition, like hangover, when it could sign any similar document without a moment's hesitation—just for being offered even in that condition a seat at a table with decent people. Earlier, Russia signed the Freon agreement without ever thinking about its chemical industry. And now it signed to stifle its not just chemical but entire industry. And energy as well.

And just recently, Russia (in the person of Putin) ratified the Kyoto Protocol signed in a hangover state (in spite of president being warned by Andrey Illarionov about the dangers of this step). Obviously, sleek western diplomats twisted arms of martial-art specialist Putin. Or promised something—who knows . . .

But it's useless to twist America's arms. Having signed the idiotic document in a state of panic delirium, America cooled down and thought hard: why the heck should we kill our own economy? Scientific paradigms change, while the economy stays. Especially that the majority of world countries did not sign the Protocol and would continue gassing in the same old way and successfully developing their economies. However, the ecologically-liberally-politically-correct European public is so much frightened by vituperation of "scientists" using the public's money, that it attributes any current weather event to climate. Any large flood, avalanche, European heat, abundant snowfall, an extra typhoon—everything is allegedly caused by global warming. Although in the great majority of cases, they were not caused by global warming but rather by normal weather fluctuations. As one of the Hollywood heroes once said, "Crap happens."

Similar weather (not climate) anomalies are known and registered in long climate series. The long climate series are those that exceed the period of contemporary instrumental observations. Only forty meteorological stations are in the whole world, which have observation series of two hundred years or more. Thus, the instrumental meteorology practically knows nothing about events that happened on most part of Earth even just a hundred years ago. And so expressions used by TV meteorologists, "unprecedented flood," "unprecedented drought," "unprecedented snowfall" are simply untrue. All this did happen, and it did happen on a much larger scale. Only you have to ask for this information not of meteorologists but of paleoclimatologists.

In fact, weather catastrophes are as normal, as normal conditions are, but just the extremes happen rarer. Calculations show that total annual damage from all natural catastrophes (including earthquakes, floods, hurricanes, eruptions) amounts to $30 billion. This number keeps growing. But it grows not because the number of catastrophes grows but because the population grows, along with the growth in number of structures, roads, power lines . . . So it's the opposite! A different trend is noted in the world—the number of natural calamities is decreasing with global warming! What grows—is the dissemination of information

—

about them. What used to happen in uninhabited areas now breaks whole settlements to pieces.

It's interesting to note that the leader and initiator of Kyoto Protocol is Germany. At first glance, it's strange, because Germany is a powerful industrial nation—metallurgy, Ruhr coalfield, developed industry . . . On the other hand, there's nothing strange here. Germans are pedantic people, obsessed with cleanness, and Germany always suffered from lots of industrial contaminants that were not only its own. The point is that prevailing winds in Europe are from the west. Everything emitted by industry in England and France falls out in Germany and almost never other way round. That's why it was Germany to always have lots of problems with acid rains, sulfur emissions . . . Accordingly, it should not be surprising that Germany was the country where a whole generation of environmental fighters grew up, who, of course, could not miss such new-fashion things like global warming and fight against greenhouse gas emissions.

In late 1980s, twenty or so not-too-well educated scientists appeared in Germany, who became members of a commission created by German Bundestag. They easily persuaded totally uneducated deputies that the humankind is about to encounter a global ecological catastrophe if urgent measures were not adopted to reduce greenhouse gas emissions. Germans even pledged to exceed the commitment and to reduce emissions by 5 percent. It was easy for Germany to fulfill its obligations: after reuniting with East Germany, the West Germans closed many East German unprofitable companies. And the East German industry used mainly lignite that produced lots of emissions.

But it should be said in all fairness that Germany did not fully surrender to crazy ecologists. Industrialists formed an opposition to Kyoto decisions, and the German Union of Coal Manufacturers played the first fiddle. Coal manufacturers and consumers were not in raptures over the fact that emissions had to be reduced, for coal has the highest coefficient of carbon emissions into the atmosphere. For example, when burning coal, emission of CO_2 is almost twice compared to burning the same amount of natural gas. That's why the coal lobby financially supports climate studies confirming that there's no need in reducing greenhouse gas emissions.

Certain scientists act like prostitutes—they will tell you whatever you pay them for it. If a scientist works for the coal lobby, he will be persuading you that emissions need not be cut. But if a scientist works for the nuclear lobby, he will try to persuade you that emissions ought to be cut, because nuclear power plants emit practically nothing and would be happy to bring down their competitors, like thermal power plants. Therefore, a scientist should work as the destitute Klimenko works in his laboratory—for love of science . . .

There is a suspicion that Kyoto Protocol ratification was greatly facilitated by the love of my compatriots to freebies. The point is that the Protocol stipulates greenhouse gas emission quotas for each country. Those countries that would not fully use their quotas could sell their "under-emissions" to the countries with developed economies, or in other words, to those who "over-emit." So here's a paradox: those who don't do a damn thing could get money from those who work. It's very fair! It's very socialistic! And it's in absolute agreement with the course of the current western political correctness and liberal idiocy.

But would we have a life full of freebies? Or are there some snags, after all?

Yes, there are. When quotas were assigned, Russia was extremely lucky to have 1990 as the reference point, which was a relatively successful year for Russian economy and power engineering. Russia was then at the peak of its emissions and was producing lots of tanks, walking excavators, and pig iron per capita. Since then, production dropped. Thus, without lifting a finger Russia can easily fulfill its Kyoto obligations and even trade its quotas. Today Russia is emitting about 150 million tons of carbon less than in 1990 and is not going to reach its 1990 level before 2020s. But later, as Russia economy develops, it will face big problems . . .

It seems strange—what kind of problems? Why do we generally need to develop some kind of economy, when we could even not lift a finger and trade quotas? However, in order to get money, someone would have to pay it. America will not pay—it withdrew from the Protocol. So we should look for money from those countries that don't want to cut their emissions. There are only three of those—Japan, Canada, and Australia. But they don't need too many quotas. Besides, Australia, upon some thinking, also withdrew from the Protocol. Russia indeed could sell quite a lot—everything that we could save during the so-called accounting period of the Kyoto Protocol (from 2008 through 2012), which is about 1 billion tons of pure carbon. But the world is ready to buy no more than 400 million tons.

Question: what's today's cost of a ton of "under-emissions"?

Answer: no one offers today more than $30 per ton. But these "under-emissions" would be sold not just by Russia. There are many sellers besides us! Ukraine, for example, has failed in its economy so badly that it could beat down the price to just a few dollars per ton . . .

But the most interesting stuff will start after 2012! By that time, the signatory countries would have to cut their emissions by 5 percent, while it's clear even now that this will never happen. And then, from bad to worse: in accordance with the scientific scheme assumed as the basis of the Kyoto decisions, the 5 percent cut in emissions that European countries were ordered would not be enough to prevent a catastrophe. By mid-twenty-first century, the 50 percent reduction in emissions would be required! And Russia would in no way be able to endure that: up to a half of all its obtained energy goes for heating! And the Kyoto Protocol never took into account the climate difference between countries. Go freeze, Russians! Or have no economy at all.

Well, in the early action of this useless Kyoto attraction, Russia will receive some small dividends, but ten years later, when the economy would be on the rise, we would ourselves have to pay for excessive emissions, otherwise our country would come up against an economic dead end.

But there are some options. For example, Russia will be getting freebie billions in the first ten years, and then it can simply withdraw from the treaty by saying, "Oh, our scientific paradigm changed! We have read some ten-year-old books, which say that there would be no catastrophic temperature increase or worldwide flood, and generally—we're facing an ice age! We're withdrawing from the Kyoto Protocol. Thank you all, good buy."

The question only remains—would Russia be allowed to act this way? If even tough Putin's arms were twisted, forcing him to sign a decision disastrous for the country, then all the more so would Russia not be let go away from the gambling table with the prize money. Russia could be hurt very badly. It's only the powerful house called the USA is allowed to act this way. When America decided the Antimissile Treaty did not benefit it anymore—it withdrew from it! When it decided the Kyoto Protocol was of no benefit to it—it refused to ratify it. The Bush administration understood perfectly that while there are numerous scientific theories, there's only one homeland, and only the scientific theory that's beneficial to your country should be adopted as the guiding one. And the Kyoto treaty is of no benefit to the States. Carbon emissions in the United States exceed one billion tons! A cut of only by 5 percent leads to losses of billions of dollars. So why do it if the scientific paradigm could be changed at no cost at all? In other words, select a group of scientists who say that nothing terrible would happen. And that on the contrary, the emissions should be increased!

Why increase them?

We'll answer, be patient. But let's first get an ultimate understanding of this freaking Kyoto Protocol.

We saw already that Kyoto Protocol is harmful and especially for Russia. But it's also simply useless for the rest of world!

Firstly, because countries that signed it emit into the atmosphere only one third of all planet's industrial emissions. Such fast-developing nations as India and China don't even think of signing it! And they have great reasoning: "Western nations have emitted so much greenhouse gases, that the planet's climate is allegedly changing. But that's how they became developed nations. So give us now a chance to become developed. And when we become developed as you are, then we'll be cutting the emissions jointly. As it is, YOU spoiled the climate and atmosphere and now want so WE would cut the emissions. Go to hell . . ." A logical address.

So, the Kyoto Protocol is useless because it closes only "one third of a hole," while the non-aligned states will blow their carbon through the other two thirds. So what's then the purpose for the signatories to suffer and stifle their economies when the rest would be developing theirs?

And now—secondly. Secondly, the Kyoto Protocol is useless because it has flawed scientific forecasts at its basis. Not potentially flawed, as you, perhaps, might think, but ALREADY proven to be erroneous.

Kyoto decisions are based on the version of catastrophic increase of global annual average temperature by 5-7 degrees (9-13°F)—but *By What Time*? And how much is 5-7 degrees (9-13°F)? For comparison, temperature increase from its lowest point in the Small Ice Age (late nineteenth century) up to the warmest decade of the second millennium (1990s) was only 0.7°C (1.3°F). An increase ten times greater could be imagined only in a state of delirium. But those are emotions. Here are the facts.

According to calculations done by "modelers," the globe should have been heated up by 3.3°C (5.9°F) since the start of the industrial era. But it has heated up since then by only 0.7°C (1.3°F). The error is almost fivefold! Yet no one committed suicide. Quite the opposite, they continue to enjoy salmon sandwiches and to frighten taxpayers at their own expense ...

An ultimately confused reader can ask, "But do these greenhouse gases generally affect global warming somehow? I read somewhere that they don't ..."

Indeed, there is a version that they don't. If you look at the global warming graphs for the last several hundred thousand years and superimpose them on the graphs of CO_2 content in the atmosphere, they correlate perfectly (fig. 13).

The upper curve here is the change of carbonic acid content in the atmosphere, and the bottom curve shows the same for methane. The middle curve is the temperature. Isn't it a perfect match? But what is the cause and what is the consequence?

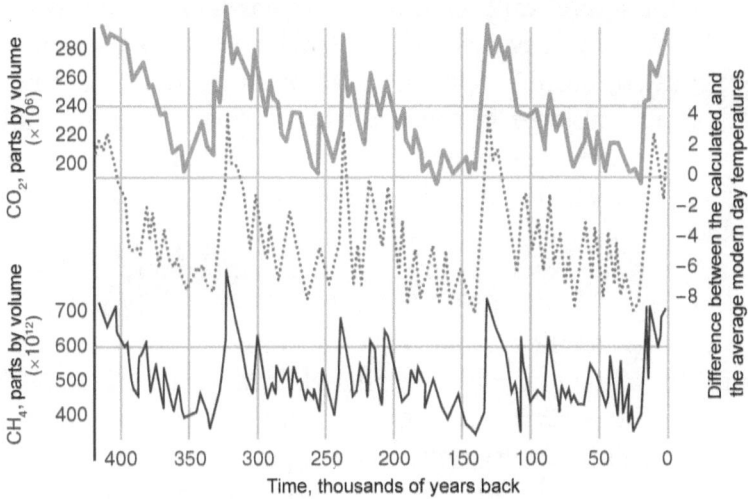

Fig. 13. Correlation of global warming periods and amounts of CH_4 and CO_2 in the atmosphere

A thorough study of the issue shows that the carbon dioxide content in the atmosphere increases AFTER the global temperature starts rising. That is, increase of CO_2 content is not the cause but rather the consequence of global warming. Where does it come from? It's from the oceans. It's known that solubility of gases in water falls with the rise in temperature. Take a glass of champagne and warm it up in your hands—it starts gassing actively. Oceans are gassing exactly the same way under global warming. And accordingly, CO_2 content in the atmosphere increases.

However, not everything is that simple. The increased content of greenhouse gases can be not only the consequence, but also the cause of a temperature increase, because no one has yet abolished the greenhouse effect. And the composition of the Earth's atmosphere has changed substantially in the last two hundred years. In the late eighteenth century, right at the start of the industrial revolution, content of carbon dioxide in the atmosphere was 280 volume parts per million, and now, it's 380. Increase by 35 percent. Even more significant was the change in methane content—it increased threefold, or by 200 percent. And one molecule of methane causes greenhouse effect twenty times greater than one molecule of carbon dioxide. Greenhouse gas contents like these were not seen on our planet for several million years. A similar CO_2 content was noted back in the Mesozoic era when the planet was literally swarmed with life. Generally, by burning coal and oil, humans return to the atmosphere the carbon accumulated in the fossil fuels, thus improving biological productivity of the atmosphere! And what's bad about rich crops and rapid growth of forests? Also, the rise of temperature and moisture. All comes together . . .

And here's the result. Annual average temperature in Moscow in 1901-1930 was +3.8°C (38.8°F) and in 1971-2000 already +5.3°C (41.5°F). Try to feel the difference! We'll remind you of one fact brought up earlier. In Russia under Nicholas I, publication of a magazine *Journal of Ministry of the Interior* started, from which you can find out that in the early twentieth century it was snowing in Moscow in June (!)—and it wasn't an exceptional case. Snow fell in June even in Kiev province.

Cold winter—more snow on the roads. Moscow newspapers once published this figure: the city spends every year up to $70 million for snow removal. And how much for the whole country? God knows—we have no such data. But we have data for a northern country Canada—they spend annually $8 billion for snow removal.

But the important thing is not snow, of course. It's the cold itself. As we know, Russia spends 50 percent of all its energy for heating residential and business buildings. Russia consumes annually from 800 to 900 million tons of equivalent fuel, depending on winter severity. The cost of 1 ton of equivalent fuel is about $300 now and is rising fast. The difference between a warm and cold winter could mean today a gain or a loss of $30 billion.

But that's not all! Fuel consumption is included in the prime cost of every product, like for example metal, called "bread of industry." And by the way, in the cold climate the need in metal is higher than in the warm one. Firstly, because steel has a feature to embrittle in the cold. And secondly, metallurgical plants of Russia spend almost 10 percent of their energy for heating up the rail vans with coal that has been frozen up during transportation. Can you imagine what that rail van with coal looks like after being transported across half the country? It's a monolith! So it's heated with hot steam. To do that, you have to boil tons and tons of water that has high heat capacity. So, it's deep butt all around. We don't need cold!

You might remember, I wrote earlier that in order to improve quality of life for Russians up to the level of developed countries, the per capita energy consumption should be increased twofold. And this would equal to having a second oil-gas field like the one in West Siberia. But we don't have a second Siberia . . .

And now we have it!

Each degree (1.8°F) of warming saves our country about 120 million tons of equivalent fuel a year. When everything is appropriately calculated, the warming in this century alone could save the country about 10 billion tons of fuel—it's much more than all the oil reserves of Russia! And all this richness comes to us right from heaven. And we meanwhile sign the Kyoto Protocol to join the entire civilized world in the brave fight against improving the quality of life for Russians.

Ladies and gentlemen! Let's fight the greenhouse gases! Content of carbon acid in atmosphere grew one and a half times! Yes, it's true. But this is a pleasant truth. Do you know that the last time such high content of carbon acid in atmosphere was observed three million years ago? And our planet looked much different then! There was no ice in the Arctic Ocean. There was no tundra, and African rhinoceros and lions inhabited northern Black Sea area, which is in today's Russia. It's not bad, is it?

Speaking of ice . . . The melting of Arctic Ocean ice is one of strategic advantages of warming for Russia. It means that the Northeast Passage would be accessible for sailing not just several weeks a year, as it is today, but many months. That means Russia could act as a real transportation bridge between the most dynamically developing regions—United Europe and East Asia. For comparison, the distance between London and Yokohama is 21,000 km (13,000 miles) via the Suez Canal and 14,000 km (8,700 miles) via the Northeast Passage.

Russia would finally become an oceanic power and not just a sea power as it was before. In its entire history, Russia had access to inland seas only—the Baltic Sea, the Black Sea, the Sea of Japan. Any access of Russia to oceans is indirect, via foreign straits, and can be easily blocked, as it did happen more than once in the past. And Russia got involved in World War I essentially for those damned the Bosporus and the Dardanelles. And the access to the Pacific Ocean is achieved only via the inland Sea of Japan. As to the Baltic Sea, in order to get into the Atlantic we have to ask for passage through Danish straits. The only ocean that Russia has direct access to is the Arctic Ocean, which is practically useless for sailing. Well, soon it will lose half its name (*in Russian it's called "Northern Ice" Ocean—Translator's note*) and would become just "Northern." And by the way, river sailing in South Russia would become all-year-round, too, and in Central Russia it would be interrupted only for several weeks a year.

So what do you think, should we fight the warming?

Chapter 7

What Will Be on Planet Earth over Next One Hundred Years?

(The chapter title is a quote from a popular Soviet Song—Translator's note.)

If we were to abstract ourselves from the catastrophic (not for Russia) predictions of the "modelers" scaring everyone with the rise of global average temperature by 5°C (9°F), if we were to forget about these fabricated scares that are *already* not coming true, and if we were to use the model that has fully proven its adequacy and forecasting power, then what do we get?

Well, it's actually this (fig. 14).

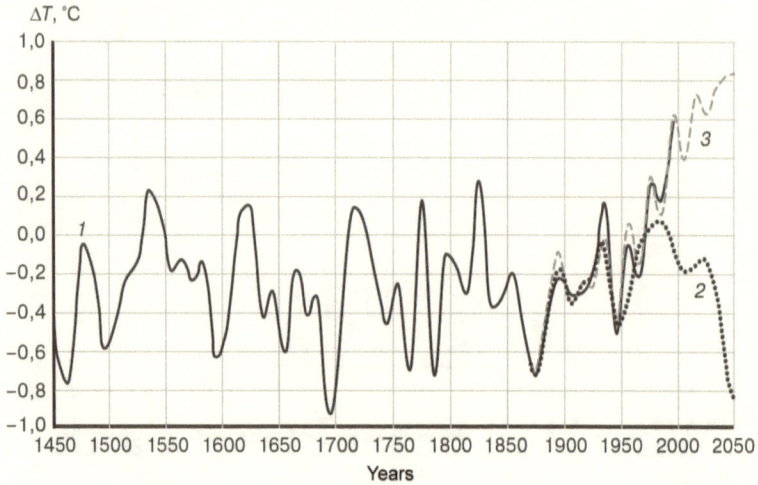

Fig. 14. History and forecast of average annual
temperature fluctuation on the territory of the
Russian plain with and without the effects of the
antropogenous factors: (1) average annual temperature
for the center of the Russian plain; (2) forecast without
the antropogenous factors; (3) forecast with the
antropogenous factors

Three curves. One is the history of temperature in the past, while
two others in the future. The latter two were obtained the following way.
Starting with 1875 (start of the industrial era), Klimenko mathematical
model was launched into the future in two versions—with and without the
anthropogenic factor. That is, what would be if the industrial revolution
never happened and humanity continued living in the bosom of bucolic
civilization? And what would be now?

As you see, the curves 2 and 3 cardinally diverge. The fate of
humankind would have been sad without the industrial revolution: if
humans were not to build industrial and power plants, were not to burn
so much fossil fuel and thus were not to raise the greenhouse gas content,
then starting with 1980s the planet would be falling into an ice age. With
all the ensuing consequences deadly for civilization. Having "gassed" the
atmosphere with carbon acid and methane, humans triggered the planet
warming. That is, their salvation.

Climate today is already warmer than at any moment in the last 3 thousand years. According to forecast, the doubling of greenhouse gases concentration on Earth will happen in the latter half of the twenty-first century, and then, due to certain geophysical reasons, it will start slowly decreasing. The maximum rise of global average temperature compared with the preindustrial era will be 2°C (3.6°F). And with an account for thermal inertia, this increase will be reached only by the late twenty-second century. It's well seen on the next graph (fig. 15).

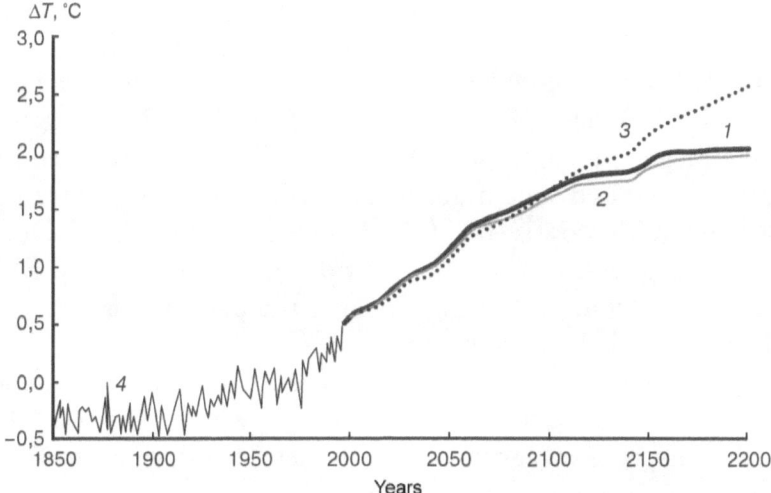

Fig. 15. Scenarios for increasing average global temperatures in the twenty-first to twenty-second centuries:
(1) base (most likely); (2) same, but with intensive forests recovery; (3) carbon emissions control, but eliminating tropical rainforests; (4) instrumental data

We have been terribly fortunate: Russia is at the center of global warming, which means that in Russia warming has a feature of intensification: 2°C (3.6°F) rise of global temperature will turn into 4-5°C (7-9°F) rise for Russia, and in some places even 10°C (18°F)! And the maximum warming will "fall" onto those places where it's especially needed—Siberia. Summer will not become hotter, but it will start a month earlier and end later. The major warming will happen in our most troublesome season—winter. That's good luck. The second good luck is the increase of moisture, which will increase the crop capacity. Of course, some regions of Russia will suffer from droughts—South Siberia, North Caucasus, parts of Central Russia. Ukraine will become severely dry. But 90 percent of Russian territory will have much more precipitation! Besides, the warming will shift the zone of profitable agriculture several hundred kilometers (few hundred miles) north. The sowing period will start earlier, providing more time for better land cultivation.

As to the grapes that already bloom in Moscow but don't fruit yet, the botanists say that grapes would need just a couple of degrees (4°F) of warmth more to set the fruit. So in about two hundred years, who knows, Moscow wine-makers would supply their wines to Siberia. And in the Black Sea area, with God's help, olives will start growing. It doesn't mean that subtropics will come to Moscow—no, such life in clover is not promised to us, unfortunately. While now the average winter temperature in Moscow is -8°C (+18°F), it will be -4°C (25°F). Don't be scared: the grape vine can endure brief frosts down to -20°C (-4°F). The most important thing is, during ripening the daily average minimum temperature should be no lower than +10°C (50°F).

However, it should be said in all fairness that even in two hundred years, it still would be too cold for grape vines near Moscow, and in these conditions the sugar synthesis in berries slows down. It means that Moscow wines would not be too good.

The highest rate of warming will be observed in the next 50 years—the temperature will rise in this half century as much as it has risen in the previous 120 years. After that, the rate of warming will slow down, and the temperature will rise by another 0.7-0.8°C (1.3-1.4°F) in the next 150 years, reaching its gentle peak at around 2200. But it will be warmer not everywhere. In most parts of the planet, it will get warmer. But in Greenland, part of China, Tibet, Himalayas, England, and eastern Mediterranean area it will get slightly cooler at first. And not everyone will enjoy higher precipitation. It will be a very tough time for Africans living in the arid zone south of Sahara. They will swell from famine and knife one another. However, I haven't seen a single Russian who would be concerned about problems of African blacks. For some reason, those problems concern mostly Hollywood actors and some of western Bohemia, who like to snuff cocaine, have exotic pets, and adopt colored kids.

Russia will have its own problem, used as a scare tool by warming fighters: the permafrost thawing. It would seem permafrost thawing is wonderful! Two thirds of Russia territory is permafrost unsuitable for life of a white person. And if the "frozen-up" part of Russia were to become normal, it would mean increase of Russia effective territory. We would not only get from heaven freebie energy that we've been lacking so badly to become a truly civilized nation with high quality of life, but we would also get some freebie territory, too!

It's all true. Then why the scare? Because everything built on permafrost would float—structures, roads, power lines ... Part of what was permafrost will become swamp, and swamp means filth ... Is it scary?

Not at all!

We've been living for a long time now under conditions of permafrost thaw. The entire West Siberia is one contiguous bog that's none other than remnants of permafrost that existed here fifteen thousand years ago. In the late nineteenth century, people in the Archangel province used to build food warehouses in permafrost. Today even traces of permafrost can't be found and local residents never heard about permafrost being there just recently.

Indeed, permafrost thaw turns area into swamp. Could construction be done on swamps? Yes, if needed. People learned to do it since two thousand years ago: ancient Romans used to build wonderful roads in German woods, which they used to transport heavy military equipment. It turned out just fine.

Only recently, Americans built a 1,000-kilometer long (620-mile) trans-Alaskan pipeline in conditions of degrading permafrost. It has floating supports to prevent the pipeline from rupturing under any movements of the ground. The cost of construction was almost twice the standard on hard ground.

Anyhow, construction in conditions of thawing permafrost is a solvable problem and not too complicated. And the process of permafrost thawing would leave people so much time for adaptation that it's even a pity—it's better to do away with it sooner. Ice doesn't melt instantaneously with temperature rising above freezing. Get some ice cubes from the refrigerator—it takes quite some time for them to melt even at room temperature. Because the heat of water phase transfer is very high! It would take hundreds years to thaw our permafrost.

—

And unfortunately, it would be destroyed not everywhere. On Taymyr and Yamal peninsulas, for instance, permafrost would not be affected at all, because it starts degrading only when the annual average temperature rises above -2°C (28°F), and the warming would never reach such highs in the northernmost areas of Russia, alas. But the destruction of permafrost during the twenty-first century would happen on the territory of 2.5 million km² (one million square miles). At least we'll get this—still it's something . . .

Long live thermal power plants, heavy industry, and millions of cars! We love it!

Vladimir Klimenko likes to repeat that the humankind has no experience living under conditions that would occur on the planet in the next two centuries and even next decades. I would add comfortable conditions.

In 1991, tourists found a dead human body in the Otztal Alps at an altitude of 3,280 meters (11,000 feet) above sea level. The body was so well preserved that the tourists first thought this was a victim of a crime and called the police. But police took a thorough look and called the scientists. The latter were in absolute raptures. It turned out that the person died accidentally, being frozen to death during his night sleep 600 years before the Egyptian pyramids were built, or about 5,300 years ago, in the Golden Age. The glacier preserved perfectly the clothing, weapon, body tissues, and even the food that the person took along . . . It means that such high temperatures as were then in that particular place have never been observed in the course of nearly five and a half thousand years—otherwise, the body would have decomposed long ago.

We are now at a threshold of a similar golden age. And for some reason, I'm not scared at all.

CHAPTER 8

After Paradise

What if we try to look further down the road, beyond the twenty-second century—what would we see there?

Nothing good. By that time, fossil fuel most likely would be almost completely burned out and the carbon emitted into atmosphere would be partially consumed by multiplied biota and insatiable ocean. The greenhouse curtain that temporarily protected us from the advance of the ice age would disappear and the temperature would inexorably slide down. All factors of nature are against us by now. And only we are still fighting for our own selves by desperately gassing carbon dioxide into the atmosphere. The graph of temperature fluctuations in the last four hundred thousand years (fig. 1) and the graph of forecasted solar activity (fig. 16) show this vividly: what we see ahead is the sharp drop of the global average temperature.

Paleoclimatology data tell us the same thing: 90 percent of the last two million years, the Earth was in the heavy embrace of ice ages. And in the course of the last three million years, the Earth climate demonstrates a steady and long-term cooling trend.

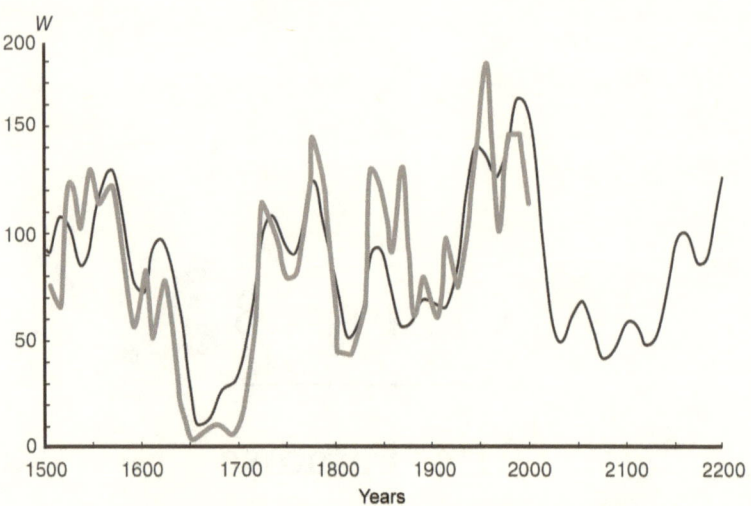

Fig. 16. History and forecast of Wolf's maximums

But there's no sense talking today about what's going to happen in hundreds years from now—this problem would be solved by our distant descendants at their science and technology level. For us to grieve for their problems would be as stupid as it would be for stone-age savages to grieve for depletion of obsidian deposits: "Oh, from what our descendants will make their knives?"

It's more important today to understand the nearest future. Because it affects the distant future.

Only the highest level of knowledge development could help civilization overcome the next ice age. Only the scientific and technological progress would enable the intellect to survive on this planet—either by artificially correcting the planet's climate in the direction of warming or by changing the structure of civilization and its bearer in such a way, that climate would generally stop affecting us.

The only problem is that the exponential growth of Earth's civilization, both scientific and demographic, happened in the last two-three hundred years, which were cold. Not deadly cold, as it was in the ice ages, but "cool"—it was not for nothing this period was called Small Ice Age. This is how it looks in numbers: temperature drop by a half or a whole degree (1-2°F) stimulates progress, inventive and creative activities. But temperature drop by eight degrees (15°F)—kills.

The golden age, which is already starting and is going to last until the next great glaciation, will be an era of climatically optimal temperatures. And times like these could be justifiably called eras of hard times—creative stagnation, as well as stagnation in all human potencies. Smooth time without embellishment. Time with no history. And Fukuyama has nothing to do here.

In general, everything that's happening on our planet seems very suspiciously strange. All turns out very well, but it's always in the very last moment, leaving no chance for a repeat. Every time we kind a squeezing in through the doors that are already closing. So far, we've been making it . . .

Humankind started the industrialization era very timely and thus saved the planet and itself from being frozen up, its efforts having moved back the ice age by fifteen hundred years.

Humankind has arrived at the stage of depleting its known fossil energy resources at the exact moment when the geological system has formed a new resource that seemed specially designed for it, which I'm not going to write about in this book because I wrote about it in other books (metal-hydride theory of Earth). I would just like to mention strange words said by geologist Vladimir Larin: "It seems everything is arranged specially for us. If I were to visit the Tunka Depression or other similar places fifteen thousand years ago, I would have not seen there anything of what's there today. But in geological terms, fifteen thousand years is nothing! It's less than an instant. Hundred millions of years—that's the scale of geological changes. That is, the planet is ripe for providing us with this resource in the very exact moment when we badly need it. An impression appears that someone specially arranges things for us to protect us from collapsing and to give us enough time to jump onto a new ice floe when the old one is crushing. I sense it in my own life—as if I'm being led to somewhere and saved from various troubles. Do you know that someone shot at me? I carry the bullet in my pocket up to now. Or another semi-mythical thing happened to me, which I still cannot solve . . ." And he told me a strange story indeed that I'm not sharing here because it better suits *Russian X-Files* book rather than a book on climate.

All this indeed reminds of some computer game with ever accelerating speed. At every crucial turn, the game's author honestly provides the player with a lifesaving platform for the next jump. And as to whether the player has time to make that jump—depends entirely on the player.

The answer to the question whether we have overcome the next barrier depends on what would be the state of civilization at the time of the Golden Age. Would it be moving ahead technologically or marking time being content with former achievements? Would there be expansion or not? Would there be a slight hunger urging to move forward or a relaxing repletion. Water never flows under a lying rock. If a person is not hit with a stick or enticed with a carrot—that person would not stir. Only the need forces to jump and think. An extended propitious period is dangerous to civilization. When blood stops moving—it might cause atrophy.

. . . Don't sleep—you'll freeze! . . .

However, there is a danger. But on the other hand, I absolutely don't want the stuff that's so adored by national-patriots and inherent to cold eras of expansionist imperialization, casualties, heroism—in the name of . . . In the name of . . . In the name of what, by the way? Is it in the name of grandeur? And what is it? And is not the global "Empire" of the Future the ultimate grandeur? Or is it maybe in the name of good life in the future? But there it is, already visible!

Hum. Indeed, why all the way in history emperors, heroes, and civilizers sacrificed their lives? For the sole reason—to help their compatriots live better. The eternal refrain: if not we, then at least our children will enjoy life! In other words, all these World War II heroes, pioneers, and other great heroes with pure soul, who are the source of pride for patriots and other citizens, are not just self-valued—their sacrifices are not senseless only because of their dedication to something or to someone. To what and to whom? To their compatriots! To their happy, pleasant, peaceful, and placid life! That is, to pure consumerism and consumers—not everyone should stand at synchrophasotrons with fanatic sparks in the eyes . . . In other words, heroes live and sacrifice for average citizens. So they could eat well, drink savory, and entertain themselves. Heroes are servants of Philistines. If a hero could be revived and placed before descendants, he/she perhaps would become indignant at their petty narrow-minded intentions, chock-fullness, and satiated indifference to their ancestors' heroism. But what was that heroism needed for if not for the future replete life? It was only for that! So here it is!

But what to do then with the replete stagnation? Stagnation scares me, too.

Perhaps, cold would not be too bad to make people move a bit. But it's so undesirable! After all, we're a southern species, we love heat and we love sun. All world civilizations were born in places with the isotherm of +18°C (64°F)—Egypt, Mesopotamia, the Indus valley, Eastern China, and all the locations of great American civilizations. For humans, temperature 18-20°C (65-70°F) is most comfortable. Specialists on ergonomics have determined that with every degree (2°F) of temperature increase, labor productivity drops by 4 percent, and at 28°C (82°F), it's reduced practically by half.

There are various ways to force people to stir. It can be a race for survival, when there's a year with no crop, then the second, third, and half of the population is brought on sled to the cemetery. Or you can make humans run round like a squirrel in a cage, because they have to pay for their mortgage, car, tuition . . .

Civilization certainly needs expansion. Stagnant immobility is as dangerous to it as hypodynamia and bedsores are to a human body. In past times, expansion was pushed by overpopulation. Excessive human resources were burned in fires of imperial wars and were poured out to colonize other continents. Civilization is forged in fight—against nature and against others like itself.

But if the nature is propitious and there's no need to compete with others like itself because of abundance of space and resources on the planet, then what would move progress forward? (I hope this would not provoke objections: nature is propitious due to warm climate of the coming golden age, while abundance of space and resources is a consequence of stabilized and perhaps future-declining population. While shortage in workforce is compensated by automated processes.)

Let's assume that people are forced to stir, being chained to bank loan debts from early years through death. So, people are occupied. But how to occupy the planetary civilization that has no competitors or a single control center? Only the center is able to set forth global tasks and accumulate means to solve them . . .

Do you think civilization can't exist without hierarchy? Perhaps, civilization can't. But humans easily can at a certain level of technology development. After all, they did exist without states (and without history) for four thousand years of the former golden age.

Something similar was in Europe around the year 1000. I can't help but quote Stefan Zweig once again, who wrote about that wonderful time:

"The year 1000. The heavy, oppressive dream paralyzed the Western World. Eyes are too tired to look around; senses are too dulled to show curiosity. The human spirit is paralyzed like after a deadly disease; humankind no longer wants to know anything about the world it inhabits. And most amazing is that everything humans knew earlier is now incomprehensibly forgotten by them. They can no longer read, write, count, and even kings and emperors of the West can't put their signature on a parchment. Sciences are stagnant and have become mummies of theology; mortal's hand is no longer capable of depicting or sculpting own body. An impervious fog covered all the horizons. No one travels anymore or knows anything about other lands; people hide in their castles and cities from savage tribes invading frequently from the East. People live in tight spaces, live in the dark, live without daring—the heavy, oppressive dream paralyzed the Western World.

"Occasionally in this heavy, oppressive somnolence, a vague remembrance sparks that once the world was different—vaster, prettier, brighter, more inspired, full of events and adventures. Weren't all these countries cut through with roads, weren't the roads used for passage by Roman legions followed by lictors, protectors of order, men of the law? Didn't a man once exist by the name of Caesar who captured both Egypt and Britain; didn't triremes cross the Mediterranean Sea and reach the countries where for a long time now not a single ship dares to sail out of fear of pirates? Didn't some king Alexander once reach India—that legendary country—and return home via Persia? Weren't there any wise men in the past who knew astrology, the wise men who knew what shape the Earth has and who mastered the mystery of humankind? All this should be read about in the books. But there are no books. It would be useful to travel, to see other lands. But there are no roads. All is gone. Perhaps, all this indeed was just a dream."

The era described by Zweig was the period of the so-called Medieval Climate Optimum. It was very warm then. We are about to reach those mellow temperatures. And after what Zweig said, this does instill certain anxiety. Wouldn't the humankind's vacation turn into new Middle Ages? Would in these hothouse conditions a potential of new knowledge be developed, enabling to overcome the next ice age?

The answer to this question must be given today. We are living at a turning point between two eras. The sense of the new world already disturbs the nostrils of the most sensitive writers and thinkers.

And don't you sense yet the aroma of something singed?

Lucky ones . . .

And at the same time, all sorts of writers like Fukuyama, Toynbee, Huntington, and Nikonov write books after books about it, but you don't read them. Recently, two buoyant Swedes, Jan Soderqvist and Alexander Bard, joined our cozy little company of futurologists. They use the same words for the same thing: the new world is coming! Some call this world postindustrial. Others call it the end of history. And yet others, like the Swedes, call it the power of "netocracy." The point, however, is not in the name but in characteristics.

So what are the basic features of this nearest future? It's easy to list them when trying to find something reassuring in them.

The world of tomorrow is the world with no God, no homeland, no family, no democracy, and no stable moral norms. It is the world of victorious consumers.

The God, once a hard-necked and quite "fleshly" being, closely monitoring the implementation of strictly defined norms, with the development of science and technology first changed his residence, having "moved" from the heaven to somewhere in the outer space, and then completely withdrew beyond the limits of space and time. At the same time, he abstracted himself to the limit, turning from a homely hot-tempered lad, with whom it was always easy to negotiate, into a strange fog from beyond with no effect on airplane flights or salaries.

The disappearance of the Absolute is followed by disappearance of the single universal coordinate system, which upon referring to church calendars always enabled to determine what's good and what's diabolic. But in today's world of humanism, where a human is the most precious value and criterion of all things, every person evaluates by own self. Homosexuality—is it good or bad? It depends whom you ask, some like it . . .

A family that used to consist of a big human group including several generations, first contracted to three persons and now went altogether down to a theoretical limit—today, according to sociological norms, a family could consist of one person. For consumerist society, it may even be good, because the consumption norm for this kind of family is higher: two persons living in one family need only one washing (dishwashing, etc.) machine, while two persons living separately need two machines!

Homeland disappears because national state disappears, and the latter is primarily associated in our minds with the notion of homeland. The talk about the coming death of states has been around for a while, and it's one of the most obvious things. Well indeed, what kind of a national state can we talk about in the conditions of the global economic space pulled together into the WTO, where migration of workforce, capitals, and goods goes freely? Local self-governing—yes; local sheriffs—yes; local fire stations—absolutely . . . But national state has simply nothing to do in this global "empire"! And no one needs any global government, because there's no one whom to oppose. The general basic principles of communal life are set. Local taxes are collected. The technosphere is functioning and self-supporting.

And finally, democracy dies in the world of future. And the point is not even that democracy dies on its own, or naturally. In the twentieth century, a steady trend is observed—fewer and fewer people go on election day to polls, because there is no economic interest: no matter who is elected, nothing would change in the grand scale of things, because the system stably protects the interests of the middle class, that is consumers. But that's not the only point. Democracy is the power of majority. But what kind of majority could we talk about in an atomized world of human individualities with each representing the absolute and ultimate minority? The system as it is perfectly protects the interests of minorities. And the addition of democracy could only worsen their situation, because the majority always votes against the minority.

And who will be ruling the world strategically? Who will be making decisions on constructing a network of thermonuclear power plants or a tunnel under the Bering Strait? Who will be accumulating resources for this? And how?

On the one hand, in a decentralized and atomized world there would be no one to do it. On the other hand, if a global project is needed, lack of democracy to a certain extent may facilitate the possibility to carry it out—there's no one to ask for it. The only important thing is to properly direct financial flows circulating at stock exchanges. And money would resolve all issues with local authorities and manufacturers.

The world is too complicated even today for entrusting its fate to homemakers or to any other voting consumers, and in about a hundred years the world's complexity and inter-intricacy will increase by another order of magnitude. Therefore, strategic issues of development should be only entrusted to society's expert associations. It's only they, the self-regulating, influential expert associations, who will dictate the capital as to which projects need the investments. Above-mentioned Soderqvist and Bard view these network associations as inaccessible virtual clubs protecting themselves with passwords and registrations from garbage opinions of suckers. The opinion of these associations, their competence, and their genius are the main power resource of the postindustrial era. And it in no way is the capital, as it was under the "old regime." It is these people, who are never elected and who upon themselves are free to accept or not to accept anyone into their closed clubs, are called "netocrats" or "network rulers" by Swedish futurologists. And I'm an informal manager of the elite. These are the people, who will replace the main class of today's society—the capitalists, just as the capitalists once replaced the previous main class—the aristocratic landowners.

And what will the others, the so-called plain people, do?

Oh, it will be a happy era for them! Ordinary people will work not-too-intensely, earn a lot, and entertain themselves in full measure. These will be exemplary, healthy consumers!

Let them consume and not poke their nose into the higher spheres. Let them fly in fat-belly airbuses for their two-month-long vacations in amusement mega-parks. Let them enjoy life by TV-watching the political circus of local elections where PR managers offer the "voters" to choose among several good-looking political dolls. Because it's pluralism . . . And because in the majority of places there will be no need in a live person to serve as techno-police manager, just as an operator is not needed in today's subway—he is simply an atavism, sitting in the cabin just for public's soothing. A living example of the former king of nature.

Most importantly—this Golden Paradise must have a social mechanism to redistribute the resources of the society toward fundamental science that is the future of civilization.

And by the way, the psychology and the way of life of people of the new world—the world that is starting already today—is the topic for another book.

INDEX

www.ingramcontent.com/pod-product-compliance
Lightning Source LLC
Chambersburg PA
CBHW031821170526
45157CB00001B/136